Redox Chemistry - From Molecules to Energy Storage

Edited by Olivier Fontaine

Published in London, United Kingdom

Redox Chemistry - From Molecules to Energy Storage
http://dx.doi.org/10.5772/intechopen.98004
Edited by Olivier Fontaine

Contributors
Sergey N. Pronkin, Cuong Pham-Huu, Nina Yu. Shokina, John R. Miller, Matthew J. Bird, Graham N. Newton, Catherine L. Peake, Darren A. Walsh, Krzysztof Fic, Elzbieta Frackowiak, Paulina Bujewska, Przemyslaw Galek, Olivier Fontaine

First published in London, United Kingdom, 2022 by IntechOpen
IntechOpen is the global imprint of INTECHOPEN LIMITED, registered in England and Wales, registration number: 11086078, 5 Princes Gate Court, London, SW7 2QJ, United Kingdom

British Library Cataloguing-in-Publication Data
A catalogue record for this book is available from the British Library

Additional hard and PDF copies can be obtained from orders@intechopen.com

Redox Chemistry - From Molecules to Energy Storage
Edited by Olivier Fontaine
p. cm.
Print ISBN 978-1-80355-537-9
Online ISBN 978-1-80355-538-6
eBook (PDF) ISBN 978-1-80355-539-3

Meet the editor

Olivier Fontaine is an associate professor at Vistec Institute, Thailand, with a dual research profile in fundamental electrochemistry and chemistry of nanomaterials. After completing doctoral studies in molecular electrochemistry, he earned multidisciplinary expertise through two post-doctorate positions at the College de France and the University of St. Andrews, Scotland. Since October 2020, he has been appointed to the Institut Universitaire de France as a junior member.

Contents

Preface

There are many facets of redox reactions in chemistry, biology, and physics. It would be impossible for a single scientist to know all the varieties of redox reactions. However, in the field of electrochemistry, there are two main families: molecular electrochemistry and the electrochemistry of electrical energy conversion and storage. It is possible to use molecules in the conversion and storage of electrical energy because the molecular mechanism involved is the same. In many instances, however, the injection or removal of one electron into or from a molecule triggers drastic changes in the nuclear framework, such as bond cleavage and bond formation.

The precise control of molecules and electrochemistry allows for storing and conserving energy. In fact, the possibility of quantitative control of molecular design is a potential strategy to increase supercapacitors' energy density and power. Mastering the competition between mass transport and the redox reaction of new molecules creates opportunities in the field of redox flow batteries. This book examines the importance of redox molecules in the development of new electrical energy storage devices.

Olivier Fontaine
Molecular Electrochemistry for Energy Laboratory,
VISTEC,
Member of Institut Universitaire de France,
Rayong, Thailand

Section 1

Introduction to Molecular Chemistry for Energy

Chapter 1

Introductory Chapter: Molecules and Materials Associated with Redox Reactions

Olivier Fontaine

1. Introduction

1992 was the year chosen by the Nobel Committee to name Professor Rudolph Marcus as the winner of this prestigious prize in chemistry [1] in recognition of his passionate investment in the atomic and electron transfer field. The consequences of the discovered formalism affect the area of redox reactions in the broadest sense: from the molecule to devices that convert or store electrical energy by the redox chemistry pathway. In modern times, redox reactions continue to be of undeniable significance for the technology that deals with the storage or conversion of energy [2].

2. Basic consideration of the electron transfer

In its simplest form, the redox reaction involves an electron donor and an electron acceptor; a molecular disruption occurs when that electron jumps from the donor to the acceptor. This disruption is formalized as a solvent reorganization in a liquid phase. The probability of a redox reaction also results from the electronic coupling strength between the electron donor and acceptor. Quickly, the electron transfer is governed by the probability that the electron can move from the donor to the acceptor. This transfer probability is modulated by the energy required to reorganize the disruption associated with the displacement of an elementary charge. The equations from (1) to (3) put these two critical parameters into formalism with H_{ab} for the coupling parameter and λ for the reorganization energy [3–5].

$$k_{ET} = \kappa_{el}(H_{ab}) \bullet \upsilon_n \bullet e^{-\frac{(\Delta G^0 + \lambda)^2}{4 \bullet \lambda \bullet R \bullet T}} \tag{1}$$

where k_{ET} is electron transfer rate constant (unit in s^{-1}), κ_{el} is electronic transmission coefficient (dimensionless), υ_n is nuclear vibration frequency (unit in s^{-1}), ΔG^0 is Gibbs-free energy (unit in eV or J), λ is reorganization energy (unit in eV or J), R is gas constant (8.314 $J{\cdot}K^{-1}{\cdot}mol^{-1}$), and T is Kelvin temperature (unit in K).

$$\upsilon_{el}(H_{ab}) = \frac{4 \bullet \pi^2 \bullet {H_{ab}}^2}{h\sqrt{4 \bullet \pi \bullet \lambda \bullet k_B \bullet T}} = \frac{{H_{ab}}^2}{\hbar} \bullet \sqrt{\frac{\pi}{\lambda \bullet k_B \bullet T}} \tag{2}$$

In Eq. (2), h is Planck constant (6.626×10^{-34} J•s or 4.14×10^{-15} eV•s), $\hbar$ is the reduced Planck constant ($\hbar = h/2\pi$), λ is reorganization energy, k_B is Boltzmann constant (1.38×10^{-23} J·K^{-1}), T is Kelvin temperature, and H_{ab} is the coupling term [6].

$$\lambda = \frac{{e_0}^2}{8 \cdot \pi \cdot \varepsilon_0} \cdot \left(\frac{1}{a_0} - \frac{1}{2 \bullet d}\right) \cdot \left(\frac{1}{\varepsilon_{op}} - \frac{1}{\varepsilon_s}\right) \tag{3}$$

For the third one, a_0 is the effective reactant radius (Stokes-Einstein radius, unit in m or Å), d is the distance from the center of the reactant to the surface of the electrode (unit in m or Å), e_0 is electron charge (constant, 1.6×10^{-19} C = 1.6×10^{-19} A·s), ε_0 is vacuum permittivity (constant, 8.854×10^{-12} F·m^{-1} = 8.854×10^{-12} s^4 · A^2 ·m^{-3} · kg^{-1}), ε_{op} is solvent optical permittivity (square of the refractive index, dimensionless value), ε_S is solvent static permittivity (dimensionless value), λ is reorganization energy (unit in J = kg ·m^2 ·s^{-2}, or eV) [7].

The formalism remains the same in its general idea in more complex forms, solid phases, or involving more complex structures than organic molecules. Solid-state physics is an example of this similarity through the small polaron theory. Eq. (4) shows the work involved in moving a small polaron. The polaron reflects the electron plus its surroundings moving in a crystal structure.

$$\lambda = -\frac{{e_0}^2}{8 \cdot \pi \cdot \varepsilon_0} \cdot \left(\frac{1}{r_p}\right) \cdot \left(\frac{1}{\varepsilon_{op}} - \frac{1}{\varepsilon_s}\right) \tag{4}$$

where r_p is radius of polaron (unit in m or Å), e_0 is electron charge (constant, 1.6×10^{-19} C = 1.6×10^{-19} A·s), ε_0 is vacuum permittivity (constant, 8.854×10^{-12} F·m^{-1} = 8.854×10^{-12} s^4 · A^2 ·m^{-3} · kg^{-1}), ε_{op} is solvent optical permittivity (square of the refractive index, dimensionless value), ε_S is solvent static permittivity (dimensionless value), and λ is reorganization energy (unit in J = kg ·m^2 ·s^{-2}, or eV). In these parameters and the relations that link these parameters, the equation resembles that of the reorganization energy.

3. Diversity of the redox reaction

Therefore, it is natural that it appeared to be of common interest to gather a set of chapters regrouping the diversities of the redox reaction in this book. All the authors have presented a broad view of redox reactions in various fields. Fundamental aspects include the influence of the nature of the electrolyte on the redox potentials, the material of pseudocapacitors, the redox-active electrolytes of supercapacitors, and the redox flow battery technology. Here, we detail the unity that links these chapters.

In Chapter 1, the authors address the phenomena of redox molecules confined inside porous carbon through the concept of Redox Mediated Electrolytes in electrochemical capacitors. Redox processes are intimately connected to the field of electrochemistry. The chapter explains the fundamentals of electrochemical capacitors and offers a complete look at this technology. A particular emphasis is placed on hybrid systems, which uses the electrolytic solution's redox activity. Electrical current creation is accomplished by using charge transfer mechanisms in all electrochemical cells. Redox reactions are utilized in several processes, including charging and discharging battery packs. In addition, similar reactions may be used to enhance the operational characteristics of other energy storage devices, such as electrochemical capacitors.

Although, in theory, the energy in electrochemical capacitors is stored electrostatically (by the formation of electrical double layers), the redox reactions introduce an additional charge and improve the energy of these systems. This is because the redox reactions enhance the energy of these systems.

The alternative way between the battery and EDLC is pseudocapacitance. The reactivity of this class of materials is not apparent, and deep electrochemical investigation is necessary, as explained in Chapter 2. Pseudocapacitance is a charge storage phenomenon that occurs at the electrode/electrolyte interface and is characterized by redox transitions. The oxidation states of one or more components of an electrode and electrolyte can shift as a direct result of modulating electrode potential. The redox reaction may be restricted to the interface, or it may propagate into the bulk of the electrode material, which will significantly increase the charge capacitance of the material. The effectiveness of charge storage owing to pseudocapacitance events is primarily determined by several fundamental criteria, including the pace of the interfacial redox reaction and its ability to be reversed. In Chapter two, the authors look at how the properties of the interfacial redox reaction can affect how well the charge can be stored in pseudocapacitive materials. In particular, discussions center on the parallels and divergences that can be drawn between the methods of charge storage utilized by batteries and pseudocapacitors. The use of impedance spectroscopy to examine the pseudocapacitive behavior of electrode material is demonstrated.

Understanding the energetics of photochemical solar energy storage, organic photovoltaics, light-emitting diodes, and even photosynthesis requires understanding redox potentials, mainly as determined by cyclic voltammetry and other similar electrochemical methods. Even though none of the energy systems that were just discussed contain substantial quantities (usually 100 mM) of the supporting electrolyte necessary for electrochemical techniques to function, these prevalent methods continue to be the most prevalent ones. At the same time, the additional electrolytes frequently have considerable impacts on the energetics that are being researched, but these effects are not recognized. Despite significant attempts to utilize microelectrodes, it has not been able to use electrochemical methods to detect redox potentials without electrolytes. This is because electrolytes are necessary for the measurement. Chapter 4 will explain novel approaches that employ the process of pulse radiolysis to partially answer the question, "What is the influence of electrolytes on redox potentials?"

Chapter 4 will aboard another type of molecular chemistry used in energy storage with the case of the redox flow battery. The development of highly soluble charge carriers capable of storing several electrons during each charge cycle is required to increase the volumetric energy density of redox flow batteries beyond that of the prototypical all-vanadium system. In Chapter 4, the authors cover the design and performance of a variety of novel charge carriers for flow batteries, emphasizing those with multi-electron redox characteristics. Specifically, they will focus on those who can accept and release multiple electrons. Polyoxometalates, fullerene derivatives, metal coordination complexes, and multifunctional organic systems are some of these compounds.

These chapters show the redox reaction's impact on the energy storage field and the answers we still need to provide.

Author details

Olivier Fontaine[1,2]

1 Molecular Electrochemistry for Energy, Vistec, Thailand

2 Institut Universitaire de France, Paris, France

*Address all correspondence to: olivier.fontaine@vistec.ac.th

References

[1] Marcus RA. Electron transfer reactions in chemistry. Theory and experiment. Reviews of Modern Physics. 1993;**65**(3):599-610. DOI: 10.1103/RevModPhys.65.599

[2] Hadjipaschalis I, Poullikkas A, Efthimiou V. Overview of current and future energy storage technologies for electric power applications. Renewable and Sustainable Energy Reviews. 2009; **13**(6):1513-1522. DOI: 10.1016/j.rser.2008.09.028

[3] Lawrence JE, Fletcher T, Lindoy LP, Manolopoulos DE. On the calculation of quantum mechanical electron transfer rates. The Journal of Chemical Physics. 2019;**151**(11):114119. DOI: 10.1063/1.5116800 [Acccessed: May 26, 2022]

[4] Marcus RA. Generalization of the activated complex theory of reaction rates. I. Quantum mechanical treatment. The Journal of Chemical Physics. 1964; **41**(9):2614-2623. DOI: 10.1063/1.1726329 [Acccessed: May 26, 2022]

[5] Marcus RA. Generalization of the activated complex theory of reaction rates. II. Classical mechanical treatment. The Journal of Chemical Physics. 1964; **41**(9):2624-2633. DOI: 10.1063/1.1726330 [Acccessed: May 26, 2022]

[6] Savéant J-M. Effect of the electrode continuum of states in adiabatic and nonadiabatic outer-sphere and dissociative electron transfers. Use of cyclic voltammetry for investigating nonlinear activation-driving force laws. The Journal of Physical Chemistry B. 2002;**106**(36):9387-9395. DOI: 10.1021/jp0258006

[7] Saveant J-M. Elements of Molecular and Biomolecular Electrochemistry. Wiley-Interscience; 2019

Section 2

Supercapacitors

Chapter 2

Redox Mediated Electrolytes in Electrochemical Capacitors

Paulina Bujewska, Przemysław Galek, Elżbieta Frąckowiak and Krzysztof Fic

Abstract

Electrochemistry is strongly related to redox reactions. Charge transfer processes are used for the current generation in all electrochemical cells. Nowadays, redox reactions are still of evitable importance for energy storage/conversion technology. For instance, the charge and discharge of batteries exploit redox reactions. Moreover, these processes can also be used to improve the operating parameters of other energy storage devices like electrochemical capacitors. Although, in principle, the energy in electrochemical capacitors is stored in an electrostatic manner (by electrical double-layer formation), the redox reactions introduce an additional charge and improve the energy of these systems. This chapter presents the principles of electrochemical capacitors' operation and provides comprehensive insights into this technology with special attention focused on hybrid systems, exploiting the redox activity of the electrolytic solution.

Keywords: energy storage devices, electrochemical capacitor, redox-active electrolytes, aqueous electrolytes, organic electrolytes, ionic liquids

1. Introduction

Growing demand for energy in all possible forms (mobility, heat, electricity) induced intensive research on various energy harvesting and storage systems. Today, it is clear that fossil fuels are no longer a reasonable choice for further society development. Various reasons, such as environmental pollution, depletion of natural resources, and remarkable climate changes, stimulated intensive research on sustainable solutions for energy harvesting and storage. In this context, numerous technologies are known for sourcing the energy in a "green" manner (like photovoltaics, wind turbines, flywheels); however, this energy must be somehow stored to be further used when needed.

Electrochemical energy conversion and storage systems are one of the most common solutions used every day by almost everyone—at home, at work, or in the car. This is true that well-known Li-ion batteries allowed the world "to move" and made our lives more "mobile". Nevertheless, these are not the only ones that have been recently developed and used. Despite the high amount of energy stored in batteries, their power density is still not enough to eliminate other technologies. Furthermore, their typical redox-based charge storage mechanism makes their lifetime short (counted in thousands of cycles) and thus less resource-effective. Electrochemical

 IntechOpen

capacitors, with their high-power density and moderate energy, cyclability counted very often in millions of cycles and much safer chemistry in the cell, appear to be an interesting technology that could serve as a standalone system or greatly accompany the battery.

2. Electrochemical capacitors: definition, construction, types

Conventional capacitors are composed of two flat, non-porous plates (electrodes) separated by a dielectric material. These devices are characterized by low energy density, limiting their application [1]. In 1957, a new group of capacitors, called electrochemical capacitors (ECs), super- or ultracapacitors, emerged. It must be here pointed out that only the "electrochemical capacitor" term should be used for scientific purposes, as other names (supercapacitors, ultracapacitors, etc.) refer to commercial products. Furthermore, "electrochemical capacitors" are often confused with "electric double-layer capacitors (EDLCs)". In fact, EDLCs are always ECs; however, this term is reserved only for the systems exploiting the double-layer charging/discharging process, thus, the mechanism is entirely electrostatic, while ECs could also exploit redox-based processes in the charge storage (like hybrid systems).

Unlike conventional capacitors, in ECs, the electrodes of highly developed surfaces are used. Such electrodes allow higher capacitance to be reached and, in consequence, the energy accumulated increases while their superior power is maintained [2].

It is worth noting that besides ECs, there are many energy storage devices and their application depends on the performance parameters. Therefore, these properties, i.e., energy and power, are crucial from a practical point of view. The so-called Ragone plot (**Figure 1**) is the best way to compare various systems' performance [4].

It can be noticed that the electrochemical capacitors demonstrate the properties between conventional capacitors and batteries—the specific power is very high, however, slightly lower than in the case of "dielectric" capacitors, and the specific energy is significantly higher—but still moderate if compared with the batteries (especially

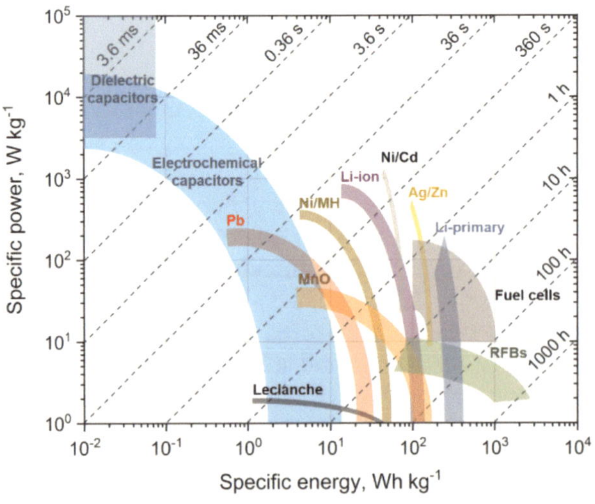

Figure 1.
Ragone plot presenting the performance parameters (energy and power) of different energy storage/conversion devices [3].

commonly used Ni/MH, Li-ion and Li-primary ones). Besides the tremendous power of ECs that allows them to be charged and discharged very quickly, these devices are getting more and more attention because of their long lifetime and safe/reliable use [5]. For these reasons, ECs are applied in the automotive industry—for instance, in regenerative braking, start-stop systems, or track control devices. Nonetheless, the energy density of these devices needs to be increased, as the volume or weight of the device must be reduced.

As already mentioned, ECs consist of two porous electrodes of highly developed surface area. The electrodes are very often made of carbon materials (especially activated carbons) due to their good conductive properties, high availability (abundance) as well as relatively low price [6]. During the ECs operation, the electrodes are polarized positively (+) and negatively (−). An insulator separates them to prevent short circuits. These components are immersed in an electrolyte, playing the role of ion source and carrier (**Figure 2a**). When charging the cell, positively charged ions (cations) are adsorbed on the (−) electrode surface, while negatively charged ions (anions) are adsorbed on the (+) electrode surface. An electrical double-layer is formed at the electrode/electrolyte interface during this process. For this reason, ECs are also called electric double-layer capacitors—EDLCs (**Figure 2b inset**). The opposite process (discharge) results in the desorption of ions from the electrode surface vicinity to the electrolyte volume (**Figure 2c**) [7, 8].

Carbon materials can be enriched with surface functional groups, heteroatoms like oxygen or nitrogen, and transition metal oxides like MnO_2. Moreover, carbon/electrically conductive polymer (e.g., PANI, PPy, PEDOT) composites can be synthesized and used as electrodes for ECs. These materials are classified as pseudocapacitive ones [9]. The charge storage mechanism in such devices can be described as quick, continuous faradaic reactions occurring with no phase change in the electrode material. The cells operating with these materials are very often called asymmetric or pseudocapacitance-based ECs (**Figure 2b**). One should restrain from using the “pseudocapacitor” term, as the pseudocapacitance concerns the electrode, not the

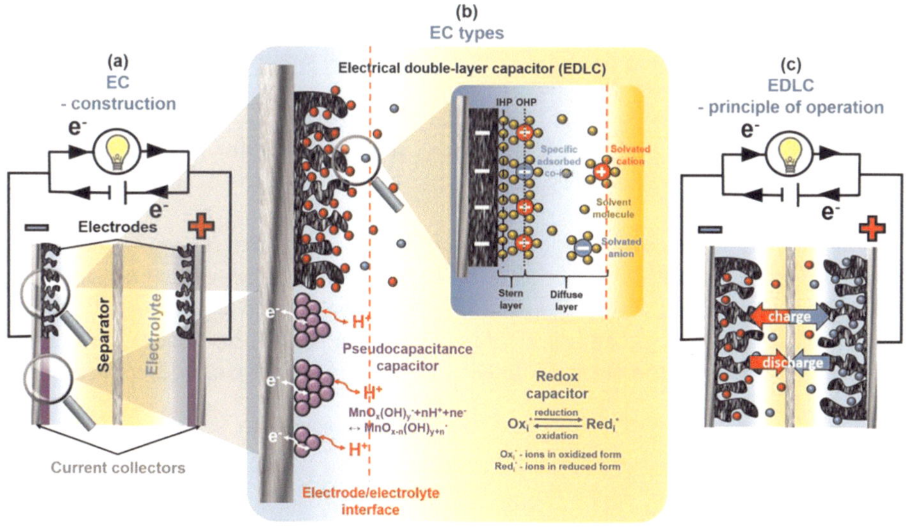

Figure 2.
Electrochemical capacitor: (a) construction, (b) types and (c) principle of operation.

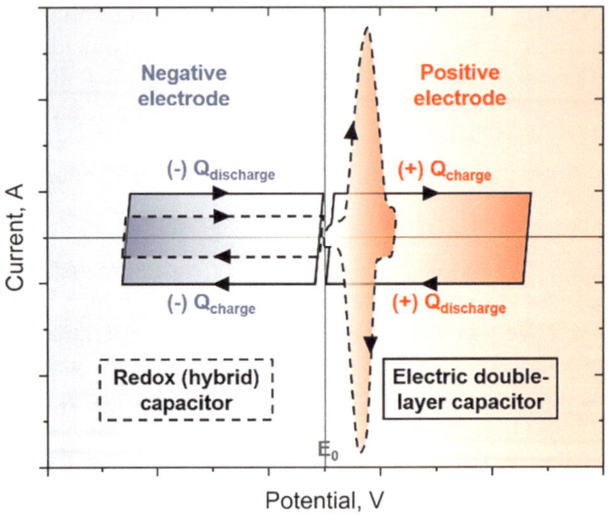

Figure 3.
Comparison of the voltammetric responses of a positively (+) and negatively (−) polarized electrode of the electric double-layer capacitor (solid line) and a redox (hybrid) capacitor (dashed line) [10].

system. If redox reactions occur on both electrodes, the system should rather be considered as a battery.

ECs incorporating pseudocapacitive materials may suffer from shorter cycle life, due to unstable behavior of the functional groups during long-term tests and chemical and mechanical composites degradation. Moreover, the cost of such materials exceeds the cost of non-modified activated carbon and impacts the final price of the cell. Thus, another solution was proposed to increase the capacitance, causing an increase in the energy of the ECs—i.e., electrolytes demonstrating redox activity (redox ECs, **Figure 2b**).

Generally, the redox processes in the batteries are attributed to the electrode material, ensuring high charge storage capacity. However, solid-state diffusion remarkably impacts the power capability. Shifting the redox processes to electrolytic solution remarkably diminishes the mass-transfer limitations and allows the power of electrostatic interactions to be almost maintained.

The operating potentials of each electrode in symmetric EDLCs are comparable. For instance, when ECs are investigated with cyclic voltammetry, the curves of rectangular shape are recorded (**Figure 3**; solid lines), since the capacitance does not depend on the potential.

In the case of galvanostatic charge/discharge, the curves are triangular [11]. Obviously, it is possible to notice potential shifts (very often negligible) that originate from matching cations/anions with the pore diameter of the electrode material. The capacitance of the system (C_{cell}) can be calculated based on Eq. (1) because two electrodes that store the energy at the electrode/electrolyte interface are considered as two capacitors in series [10].

$$\frac{1}{C_{cell}} = \frac{1}{C^{+}} + \frac{1}{C^{-}} \tag{1}$$

Assuming the capacitance values of both electrodes in symmetric cell are comparable ($C^{+} \approx C^{-} = C_{ele}$), Eq. (1) can be transformed to Eq. (2):

$$C_{cell} = \frac{C_{ele}}{2} \tag{2}$$

The specific energy for the EDLCs (E_{EDCL}) can be calculated from Eq. (3):

$$E_{EDCL} = 0.5CV^2 \tag{3}$$

where C can be calculated from Eq. (4):

$$C = \frac{Q}{m\Delta V} \tag{4}$$

For accurate calculations, it is necessary to consider the ohmic drop for ΔV calculation [10].

In the case of ECs operating in redox-active electrolytes, the potential range of each electrode can significantly differ, as presented in **Figure 3** (dashed lines). It is seen that one electrode demonstrates capacitive character, typical of EDL formation, with constant capacitive current recorded; at the same time, the positive electrode demonstrates a very high current response with a narrow potential range. This suggests high capacity, accumulated in a narrow potential range, typical of the redox process. In the galvanostatic charge/discharge technique, the redox activity is seen as a *plateau* on the $E = f(t)$ plot [11].

For the cells' performance characterization, the specific energy (E) should be calculated from the galvanostatic charge/discharge profile, with applied current I and change in the voltage (V) over the time (t), recalculated per active mass (m) of both electrodes:

$$E = \frac{1}{3600 \cdot m} \int VIdt \tag{5}$$

Power capability needs to be calculated as well for the full characterization of the investigated cells. It is directly related to the system's energy, according to Eq. (6):

$$P = \frac{E}{\Delta t_{disch}} \tag{6}$$

where Δt_{disch} is the discharge time at which the energy is released.

For more detailed information and characterization techniques, comprehensive literature reports are published [9, 12, 13].

It must be clearly stated that the energy and power of the devices should be expressed per mass of the cell components and must not be calculated for the single electrode. However, on the laboratory scale, when the electrolyte is in great excess, only the mass of the electrolyte confined in the pores should be - considered. The other possibility is to normalize these values per volume of the device's components. All the presented methods of cells characterization is correct, but the author needs to comment on how the calculations were made [10–12].

3. Redox-mediated electrolytes

As mentioned, the redox-active electrolyte in EC allows the cell performance to be significantly improved. It is necessary to use the electrodes made of electrically

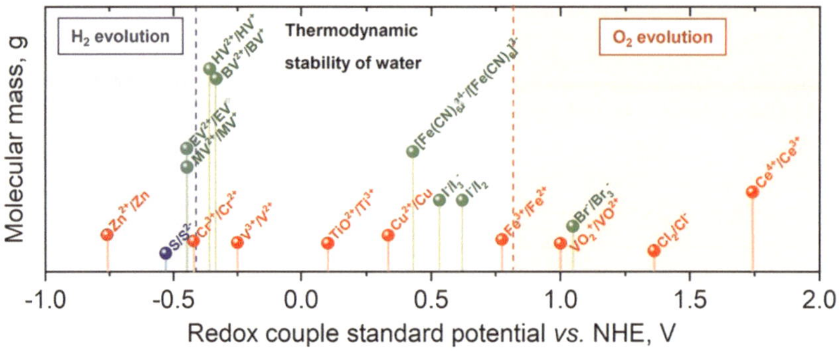

Figure 4.
Redox couples with their reduction potentials [18]. Redox couples marked in red are stable in acidic conditions, those marked in green are stable in neutral solutions and the one in blue is stable in alkaline electrolytes.

conductive material to make the electron flow from the electrode to the electrolyte possible [14–17].

There are many redox couples with well-defined and stable redox activity that can be used as additives for electrolytic solutions. The most popular ones, with their reduction potentials (expressed *vs.* normal hydrogen electrode; NHE), are presented in **Figure 4** [18].

Depending on the cell construction, electrode material used, potential application, and expected operating performance, one can select which redox couple is suitable for EC that meets the requirements. In the case of aqueous-based systems, there are additional issues that need to be taken into account. First of all, at too high or too low potentials, water is decomposed, so oxygen and hydrogen evolution can be observed, respectively. These reactions are considered harmful for the cell because (i) the solvent should not be decomposed, (ii) evolving gases can block the electrode porosity, (iii) the highly active oxygen causes the irreversible electrode oxidation and its degradation, and (iv) corrosion of the current collectors remarkably affects the cell lifetime. Therefore, the potential of the chosen redox couple should preferably be between hydrogen (HEP) and oxygen (OEP) evolution potential.

The second issue is related to the electrolyte pH. Both HEP and OEP are pH-dependent—when the solution pH increases, these potentials are shifted toward lower potentials [19]. It is, thus, possible to slightly adjust the HEP and OEP by regulating the electrolyte pH. However, one should keep in mind that the potentials of some redox couples are also pH-dependent, so with the pH change, their potential will also change. Moreover, the stability of redox couples also depends on the solution pH.

Redox-active electrolytes are grouped in a way similar to the types of electrolytes. Hence, they can be divided into two main groups—aqueous and nonaqueous ones [11, 20].

3.1 Aqueous redox-active electrolytes

Aqueous solutions, despite their limited operating voltage related to the theoretical water decomposition above 1.23 V, are very attractive electrolytes for ECs due to their price lower than for nonaqueous electrolytes and the possibility of the cell manufacturing in an ambient atmosphere. Moreover, the impact of water-based solutions on the environment is rather negligible. These solutions are also characterized by high conductivity and low viscosity. The main drawback of the ECs operating in redox-active electrolytes is moderate cycle life related to the efficiency of the redox reactions and possible side reactions [21, 22].

In general, it seems beneficial to combine more than one redox additive in one electrolyte. If the ratio between different redox species is well-optimized, the energy reached in such cells is higher than reported for the single redox couple [23, 24].

Aqueous redox-active electrolytes can be divided into three groups: cationic, anionic, and neutral electrolytes, due to the charge of the redox-active ion. It is worth mentioning that the redox ions in cationic and anionic electrolytes contribute to the EDL formation, whereas in neutral electrolytes redox species quite often do not participate in this process [11].

3.1.1 Cationic aqueous redox-active electrolytes

Cationic redox electrolytes can be divided into three groups: lanthanides, transition metals, and organic species [18, 25–31]. The general requirement is that the solubility of these species should be possibly high and their standard potential should be close to HEP, as their activity is expected at the negative electrode [18].

Cerium, which belongs to lanthanides, was introduced to the acidic solution [11, 27]. However, standard redox potentials of lanthanides (~+1.6 V vs. NHE of Ce^{3+}/Ce^{4+} redox couple) being higher than OEP definitely limits their application.

The second group—transition metals like Zn, Sn, Mn, Fe, Ni, Cu, include a solid phase in the solution of neutral or acidic pH. The cations are reduced at relatively low potential, between −0.762 V and +0.337 V vs. NHE. However, still irreversible hydrogen evolution reaction can occur during the metal electrodeposition in aqueous solutions. Although in general hydrogen evolution reaction is considered parasitic or unwanted, it is possible to store hydrogen reversibly in the electrode porosity—it is necessary to use microporous electrodes for this purpose. Moreover, the addition of halide ions to the electrolytic solution can be beneficial—halide anions will block the carbon and its active sites preventing hydrogen reactions [32, 33]. Finally, metal electrodeposited on the electrode can affect the specific surface area of the electrode and worsen the performance stability of the system. It is also possible to avoid solid-state metal deposition on the electrode, by applying redox couples dissolved in the liquid state, like Fe^{2+}/Fe^{3+}, Cu^{+}/Cu^{2+} [26, 34].

Viologen di-cations can be included in the organic cationic additives. Moreover, these species are characterized by fast redox kinetics and high reversibility [35, 36]. It was found that 1,10-dimethyl-4,40-bipyridinium cation (MV^{2+}) is strongly attracted to the electrode surface. However, after reduction to MV^{+}, the physical interaction between these species and the electrode can be even stronger [37]. This may be beneficial to reduce self-discharge, which is caused by redox shuttling.

As the cations are supposed to be attracted to the negatively polarized electrode, redox reactions originating from cationic additives are mostly at the negative side. However, the synthesis of carbon material exhibiting the affinity to cations and application of such an electrode as the positive one in ECs is also reported [26, 38–40]. It is worth noting that not only carbon materials can be functionalized—in fact, but various polymers can also be enriched with cationic (or anionic) functional groups.

The systems operating in redox-active electrolytes with transition metals as active species need to be assembled with ion-selective membranes as separators. These membranes can mitigate the self-discharge and leakage current which are relatively high for such systems [23, 39]. Nevertheless, the application of viologens (organic molecules) as a redox additive to the electrolytic solution can also decrease self-discharge without the necessity of ion-selective membrane employment. It is caused by viologens strong adsorption at the porous electrode surface [18, 28].

The main disadvantage of using viologens is their limited solubility and large size of the molecule that can negatively influence the ECs performance [32, 41], especially because of mass-transport issues.

3.1.2 Anionic aqueous redox-active electrolytes

The anionic redox-active electrolytes contain halides (iodide [14, 42–44], bromide [4, 45], pseudohalides (thiocyanate [41], selenocyanate [46]), organometallic complexes (ferricyanide and, ferrocyanide [21, 47–53]) and organic anion—like indigo carmine [54].

In the case of halide and pseudohalides-based electrolytes, a well-defined redox response is recorded at the positively polarized electrode. They are characterized by strong adsorption at the electrode surface. Hence, the self-discharge of the cell operating in such electrolytes is relatively low and the application of an ion-selective membrane is not needed. Moreover, halides can be coupled with metal ions deposition reaction, especially Zn/Zn^{2+}, and viologen redox couple [30, 31, 55], however, such systems are no longer typical capacitors. To avoid metal dendrites formation, some additional components should be used, like dendrite suppression or nanoporous separators [56–59].

The standard potentials of bromide and iodide reactions are similar; however, the bromides demonstrate slightly higher values [60]. It can be beneficial for reaching higher energy of the ECs, as the operating voltage might be shifted toward higher values. Nonetheless, bromide solutions are toxic, so for safety, it is favorable to use iodide-based solutions. Also due to the high standard potential of Br^-/Br_2, close to oxygen evolution potential, the electrolyte decomposition can be difficult to control and corrosion on current collectors can be observed [31]. Iodide-based ECs are widely described in the literature. These systems are characterized by stable operation even during long-term experiments [32, 61, 62].

Pseudohalides solutions exhibit similar electrochemical behavior to halide solutions when used as electrolytes in ECs but self-discharge is definitely more pronounced. Thiocyanates-based solutions are especially interesting for ECs application due to their higher maximum operating voltage than selenocyanate-based electrolytes. Moreover, the energy and power of such systems are comparable to those reached in iodide-based electrolytes, but their lifespan is still limited [41].

Organometallic-based electrolytes (ferricyanide- or ferrocyanide-based solutions) ensure the promising performance of the ECs. The main drawback of these electrolytes is high self-discharge, seen as low efficiency, especially at low current loads. Therefore, ion-selective membranes are very often used to limit redox shuttling [48, 63].

3.1.3 Non-ionic aqueous redox-active electrolytes

Even in aqueous-based systems, organic redox-active additives can be used. For instance, hydroquinone (HQ), anthraquinone [64–66], catechol (an isomer of benzoquinone), rutin [67], p-phenylenediamine [68], and conducting polymers [69, 70] (if soluble in water) are popular neutral electroactive species added to the electrolytic solutions. To enable redox reaction with proton transfer, the use of supporting electrolytes is necessary. For this reason, acid solutions (H_2SO_4) are used as a source of protons. As a consequence, the maximum operating voltage of the ECs operating in an acidic medium is limited to ~1 V, and, because of corrosion issues, the use of gold or other noble metal current collectors is necessary.

As the representative reaction, the reduction of benzoquinone to hydroquinone (Q/HQ) is presented in Eq. (7).

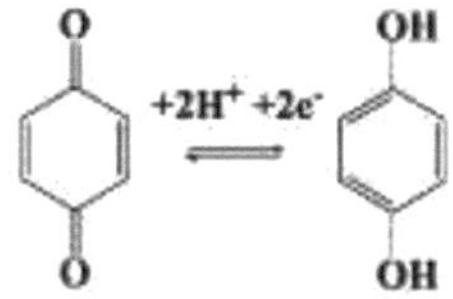

(7)

Moreover, the conductivity of the electrolytic solutions with organic molecules can be diminished. Therefore, additional ionic species are very often introduced (like neutral salts—KNO_3 or alkaline KOH [71]); hence, the formation of EDL can be more efficient. These systems are also characterized by considerable self-discharge related to the movement of neutral molecules between the polarized electrodes. To reduce self-discharge and increase the efficiency of the charging and discharging processes, the use of an expensive proton exchange membrane is recommended, which significantly increases the price of ECs [72]. The cells operating in the electrolytes with polymeric additives (i.e., sulfonated polyaniline or p-nitroaniline) also required the use of cheaper membranes. It is possible to use a semipermeable membrane that allows the movement of protons and supporting ions like SO_4^{2-}. The drawback of such electrolytes is the solubility of the polymeric molecules—when the concentration of electroactive molecules is relatively low, the capacity of the cell is also limited [69]. Therefore, it is necessary to investigate the ECs with new polymer-based electrolytes to develop these systems and reach satisfactory operating parameters.

3.1.4 Cationic-anionic electrolytes

As cationic additives exhibit redox activity at the negatively polarized electrode and anionic additives at the positively polarized one, they can be combined, giving significant performance improvement. These redox couples should be carefully selected because they must be stable and soluble under the same conditions. Otherwise, it would be necessary to use more expensive separators/membranes and the assembly process would be more complex [18]. ECs operating in the electrolyte containing viologen cation and halide anion were tested. In the case of the electrolyte with MV^{2+} and I^- redox-active species during cell charging, an irreversible capacitance loss was noticed. It was caused by precipitate formation ($MX^{\bullet+}$–I^-) [73]. When the iodide was replaced by bromide (the anion of higher standard potential) the processes were reversible, and higher energy was reached. However, because of the high potential needed for Br^-/Br_3^- activity, the signs of corrosion were observed. $MVCl_2$/KBr-based cells suffer from a relatively high self-discharge, which was more pronounced than for halide-based electrolytes, suggesting that MV species are, mostly, responsible for this voltage loss. Therefore, other viologen was used—1,10-diheptyl-4,40-bipyridinium dibromide ($HVBr_2$), resulting in lower self-discharge. Probably, not only stronger adsorption of HV^{2+} cation was the reason for the lower self-discharge but also these cations were immobilized due to the precipitate formation within the carbon electrode [74]. The optimization of redox-active species concentration, choice of the appropriate counter anion/cation for redox-active cation/

anion, respectively, and experimental conditions optimization is definitely more complex and time-consuming than for one active component within the electrolytic solution. However, taking into account the significant improvement of the energy stored in the EC operating in the redox-active electrolytes, it is still worth discovering the potential of this field.

3.2 Redox-mediated nonaqueous electrolytes

Commercially used ECs very often employ nonaqueous electrolytes (organic ones) despite the fact, that they cannot be considered environmentally friendly solutions, because of the necessity of toxic solvents use—like acetonitrile or propylene carbonate. However, they have a few advantages that make them more attractive for ECs construction: wider electrochemical window (up to 3.8 V [75, 76]) which allows higher energy to be stored, and long cycle life. On the other hand, there are ionic liquids called "green solutions", that can be also used in ECs but they are relatively expensive.

3.2.1 Organic electrolytes with redox activity

In the organic electrolytes, conductive salts like tetraethylammonium tetrafluoroborate ($TEABF_4$), lithium bis(trifluoromethanesulfonyl)imide (LiTFSI), 1-ethyl-3-methylimidazolium bis(trifluoromethylsulfonyl)imide (EMImTFSI), lithium hexafluorophosphate ($LiPF_6$), are dissolved in acetonitrile (ACN) or propylene carbonate (PC), which are the most popular solvents for ECs application [22, 77–81]. As already mentioned, organic electrolytes allow the ECs to operate at higher voltages than aqueous-based electrolytes do [82] and they provide higher power than the systems with ionic liquid (IL) electrolytes [11, 83] due to the higher ionic conductivity of organic electrolytes.

The maximum voltage of reported organic-based cells is 2.5 V, when IL (1-ethyl-3-methylimidazolium ferrocenylsulfonyl-(trifluoromethylsulfonyl)-imide, [EMIm][FcNTf]) in ACN [76] and p-phenylenediamine additive to lithium perchlorate $LiClO_4$ in ACN [84] were used as electrolytes. The mixture of microporous carbon with carbon black and graphite was used as the electrode material. However, there are also other materials that can be used, for instance two-dimensional titanium carbide (MXene) [81].

Organic electrolytes exhibiting redox activity are not as popular as aqueous electrolytes. Therefore, there is a gap in this field of study as there are many possible redox additives that could be employed for organic electrolytes [85].

3.2.2 Ionic liquids

Ionic liquids (ILs) are compounds composed entirely of ions—bulky, usually asymmetric organic cation and anion (weakly coordinating) that can be both organic and inorganic [86, 87]. As they are ionic conductive, there is no need to use additional solvents. They are characterized by high electrochemical stability, ensuring a high voltage window (>3 V) and high thermal stability [88, 89]. It is possible to introduce redox additives to IL, for example by incorporating metal ions (Cu^{2+} added in the form of copper chloride to [EMIm][BF_4] [40], neutral redox molecules [90, 91] (HQ added to [TEA][TFSI] [92]) or sulfates ($SnSO_4$ and $VOSO_4$ [23]).

However, ILs themselves can also exhibit redox activity if an anion of IL is electroactive. Hence, such an electrolyte can be called redox-active IL. To observe effective and beneficial redox contribution to ECs charge/discharge, a high concentration of electroactive species needs to be ensured. Electrolyte composed of two ILs—[EMIm][BF_4] and [EMIm]Br, where the latter one is a redox additive (1 mol L^{-1}) to the former one, was used in microporous electrodes-based EC. The operating parameters were significantly improved due to the bromide activity (the specific energy was almost twice higher if compared to the [EMIm]BF_4-based system, where only EDL formation is assumed, and the Coulombic efficiency was ~100%) [83]. Moreover, the leakage current was reduced, probably due to strong adsorption of halide on the positively charged carbon electrode, described for aqueous-based cells [18, 32].

Biredox ILs can also be used as electrolytes for ECs. The idea arises due to the potential balancing issue when additives, like metal ions, HQ, or redox-active anions, are introduced to the system operating with microporous carbon electrodes [75]. The cation of IL ([BMIm][TFSI]) was functionalized with AQ, whereas the anion was functionalized with 2,2,6,6-tetramethylpiperidinyl-1-oxyl (TEMPO) molecule. The energy density of such a system was definitely higher than for IL with redox additive as an electrolyte but the specific power and lifetime were rather moderate.

The application of redox-active IL as electrolytes in ECs is a promising strategy to increase the specific energy of the systems. However, one should take into account that the price of such devices is relatively high. Moreover, the power performance and the lifespan of the ECs operating in ILs should be improved.

3.2.3 Redox-mediated gel electrolytes

Gels are characterized by very good stability (both chemical and mechanical) and they can be made of eco-friendly materials [93]. They can be successfully applied as electrolytes (based on aqueous solutions or ILs) for ECs [61, 94, 95]. Gel electrolytes were introduced to ECs to reduce their self-discharge [96] and enable the development of flexible devices, where liquid electrolytes would expose the cells to leaks. Redox mediators can be introduced to the gel electrolytes and increase both ionic conductivity and capacitance of the ECs [97]. For instance, when indigo carmine was added to the gel electrolyte based on polyvinyl alcohol (PVA) and sulfuric acid, the ionic conductivity increased by almost 190% [98]. Moreover, the lifetime of the devices can be prolonged. Redox-active compounds, like 1-butyl-3-methylimidazolium iodide and bromide (BMImI, BMImBr) [99, 100], 1-anthraquinone sulfonic acid sodium [101], 1,4-naphthoquinone [102], including indigo carmine [98] and $FeBr_3$ [103], can be incorporated into gel structure. BMImBr with Li_2SO_4 as an additive to the PVA-based gel electrolyte was reported as a perfect solution for lowering self-discharge, increasing energy, and lifetime of the EC [100]. Flexible capacitors based on gel electrolyte—poly (methyl methacrylate)-propylene carbonate-lithium perchlorate electrolyte with HQ as a neutral redox additive were also investigated [104].

4. Summary

Redox-active electrolytes can be successfully applied in electrochemical capacitors and these electrolytes remarkably improve the energy density. It is crucial to use redox additives with a well-reversible and well-defined redox response, as the efficiency of charging/discharging should not be affected by redox process.

A variety of redox couples can be selected depending on the user's requirements: for the systems based on aqueous or nonaqueous electrolyte, with redox species supposed to be active at the positively or negatively polarized electrode, or which parameters are the most important—high energy, high power, or very long cycle life. Taking into account aqueous-based redox-active electrolytes, the most attractive from the practical point of view are cationic and anionic electroactive species—because of their good solubility in water ensuring high conductivity of the solution. Moreover, the cells operating in organic/polymer-based electrolytes are more expensive due to the proton/ions permeable membranes that have to be used.

There are also a few issues that need to be solved. Redox species cause higher self-discharge of the cell in comparison to ECs with pure EDL formation. Therefore, it would be beneficial to "trap" the redox species within the pores of the material to prevent their movement to the electrolyte bulk. Moreover, the lifetime of EC with redox-active electrolytes should be prolonged, because it is still significantly shorter than the lifetime of the cell operating in the typical capacitive electrolytes.

Nevertheless, redox-active electrolytes in electrochemical capacitors offer an interesting alternative to the solid-state compounds and composites with maintained power and improved charge/discharge efficiency.

Acknowledgements

European Research Council Starting Grant 2017 project "IMMOCAP" (GA 759603) is acknowledged for financial support covering The Open Access Publishing Fee.

Author details

Paulina Bujewska, Przemysław Galek, Elżbieta Frąckowiak and Krzysztof Fic*
Poznan University of Technology, Institute of Chemistry and Technical Electrochemistry, Poznan, Poland

*Address all correspondence to: krzysztof.fic@put.poznan.pl

References

[1] Barber P, Balasubramanian S, Anguchamy Y, Gong S, Wibowo A, Gao H, et al. Polymer composite and nanocomposite dielectric materials for pulse power energy storage. Materials. 2009;**2**(4):1697-1733

[2] Ho J, Jow TR, Boggs S. Historical introduction to capacitor technology. IEEE Electrical Insulation Magazine. 2010;**26**(1):20-25

[3] Simon P, Gogotsi Y. Materials for electrochemical capacitors. Nature Materials. 2008;**7**(11):845-854

[4] Maher M, Hassan S, Shoueir K, Yousif B, Abo-Elsoud MEA. Activated carbon electrode with promising specific capacitance based on potassium bromide redox additive electrolyte for supercapacitor application. Journal of Materials Research and Technology. 2021;**11**:1232-1244

[5] Kötz R, Carlen M. Principles and applications of electrochemical capacitors. Electrochimica Acta. 2000;**45** (15–16):2483-2498

[6] Beguin F, Frackowiak E, editors. Carbons for Electrochemical Energy Storage and Conversion Systems 1st ed. CRC Press; 2009

[7] Béguin F, Frackowiak E. Supercapacitors: Materials, Systems, and Applications. Weinheim, Germany: Wiley-VCH Verlag GmbH & Co. KGaA; 2013

[8] Simon P, Gogotsi Y. Perspectives for electrochemical capacitors and related devices. Nature Materials. 2020;**19**(11): 1151-1163

[9] Brousse T, Belanger D, Long JW. To be or not to be pseudocapacitive? Journal of the Electrochemical Society. 2015; **162**(5):A5185-A5189

[10] Gorska B, Frackowiak E, Beguin F. Redox active electrolytes in carbon/carbon electrochemical capacitors. Current Opinion in Electrochemistry. 2018;**9**:95-105

[11] Lee J, Srimuk P, Fleischmann S, Su X, Hatton TA, Presser V. Redox-electrolytes for non-flow electrochemical energy storage: A critical review and best practice. Progress in Materials Science. 2019;**101**:46-89

[12] Balducci A, Belanger D, Brousse T, Long JW, Sugimoto W. A guideline for reporting performance metrics with electrochemical capacitors: From electrode materials to full devices. Journal of the Electrochemical Society. 2017;**164**(7):A1487-A14A8

[13] Laheäär A, Przygocki P, Abbas Q, Béguin F. Appropriate methods for evaluating the efficiency and capacitive behavior of different types of supercapacitors. Electrochemistry Communications. 2015;**60**:21-25

[14] Sankar KV, Kalai SR. Improved electrochemical performances of reduced graphene oxide based supercapacitor using redox additive electrolyte. Carbon (New York). 2015; **90**:260-273

[15] Sun K, Feng E, Peng H, Ma G, Wu Y, Wang H, et al. A simple and high-performance supercapacitor based on nitrogen-doped porous carbon in redox-mediated sodium molybdate electrolyte. Electrochimica Acta. 2015;**158**:361-367

[16] Fan L-Q, Zhong J, Zhang C-Y, Wu J-H, Wei Y-L. Improving the energy

density of quasi-solid-state supercapacitors by assembling two redox-active gel electrolytes. International Journal of Hydrogen Energy. 2016;**41**(13):5725-5732

[17] Wang C, Xi Y, Wang M, Zhang C, Wang X, Yang Q, et al. Carbon-modified $Na_2Ti_3O_7 \cdot 2H_2O$ nanobelts as redox active materials for high-performance supercapacitor. Nano Energy. 2016;**28**: 115-123

[18] Chen S-E, Evanko B, Wang X, Vonlanthen D, Ji X, Stucky GD, et al. Design of aqueous redox-enhanced electrochemical capacitors with high specific energies and slow self-discharge. Nature Communications. 2015;**6**(7818): 1-10

[19] Pourbaix M. Atlas of Electrochemical Equilibria in Aqueous Solutions/By Marcel Pourbaix; Translated from the French by James A. Franklin (Except Sections I, III 5, and III 6, which were Originally Written in English). 2d English ed. Houston, Tex: National Association of Corrosion Engineers; 1974

[20] Zhang L, Yang S, Chang J, Zhao D, Wang J, Yang C, et al. A review of redox electrolytes for supercapacitors. Frontiers in Chemistry. 2020;**8**:413

[21] Chodankar NR, Dubal DP, Lokhande AC, Patil AM, Kim JH, Lokhande CD. An innovative concept of use of redox-active electrolyte in asymmetric capacitor based on MWCNTs/MnO_2 and Fe_2O_3 thin films. Scientific Reports. 2016;**6**(1):39205

[22] Singh A, Chandra A. Enhancing specific energy and power in asymmetric supercapacitors—A synergetic strategy based on the use of redox additive electrolytes. Scientific Reports. 2016; **6**(1):25793

[23] Lee J, Krüner B, Tolosa A, Sathyamoorthi S, Kim D, Choudhury S, et al. Tin/vanadium redox electrolyte for battery-like energy storage capacity combined with supercapacitor-like power handling. Energy & Environmental Science. 2016;**9**(11): 3392-3398

[24] Teng Y, Liu E, Ding R, Liu K, Liu R, Wang L, et al. Bean dregs-based activated carbon/copper ion supercapacitors. Electrochimica Acta. 2016;**194**:394-404

[25] Ren L, Zhang G, Yan Z, Kang L, Xu H, Shi F, et al. High capacitive property for supercapacitor using Fe^{3+}/Fe^{2+} redox couple additive electrolyte. Electrochimica Acta. 2017;**231**:705-712

[26] Sun XN, Hu W, Xu D, Chen XY, Cui P. Integration of redox additive in H_2SO_4 solution and the adjustment of potential windows for improving the capacitive performances of supercapacitors. Industrial & Engineering Chemistry Research. 2017; **56**(9):2433-2443

[27] Díaz P, González Z, Santamaría R, Granda M, Menéndez R, Blanco C. Enhanced energy density of carbon-based supercapacitors using cerium (III) sulphate as inorganic redox electrolyte. Electrochimica Acta. 2015;**168**:277-284

[28] Roldán S, Granda M, Menéndez R, Santamaría R, Blanco C. Supercapacitor modified with methylene blue as redox active electrolyte. Electrochimica Acta. 2012;**83**:241-246

[29] Evanko B, Boettcher SW, Yoo SJ, Stucky GD. Redox-enhanced electrochemical capacitors: Status, opportunity, and best practices for performance evaluation. ACS Energy Letters. 2017;**2**(11):2581-2590

[30] Sathyamoorthi S, Kanagaraj M, Kathiresan M, Suryanarayanan V, Velayutham D. Ethyl viologen dibromide as a novel dual redox shuttle for supercapacitors. Journal of Materials Chemistry A, Materials for Energy and Sustainability. 2016;**4**(12):4562-4569

[31] Yoo SJ, Evanko B, Wang X, Romelczyk M, Taylor A, Ji X, et al. Fundamentally addressing bromine storage through reversible solid-state confinement in porous carbon electrodes: Design of a high-performance dual-redox electrochemical capacitor. Journal of the American Chemical Society. 2017;**139**(29):9985-9993

[32] Lee J, Srimuk P, Fleischmann S, Ridder A, Zeiger M, Presser V. Nanoconfinement of redox reactions enables rapid zinc iodide energy storage with high efficiency. Journal of Materials Chemistry A. 2017;**5**(24): 12520-12527

[33] Fic K, Meller M, Frackowiak E. Interfacial redox phenomena for enhanced aqueous supercapacitors. Journal of the Electrochemical Society. 2015;**162**(5):A5140-A51A7

[34] Mai L-Q, Minhas-Khan A, Tian X, Hercule KM, Zhao Y-L, Lin X, et al. Synergistic interaction between redox-active electrolyte and binder-free functionalized carbon for ultrahigh supercapacitor performance. Nature Communications. 2013;**4**(1):2923

[35] Monk P, Mortimer R, Rosseinsky D. Electrochromism and Electrochromic Devices/Paul Monk, Roger Mortimer, David Rosseinsky. Cambridge: Cambridge University Press; 2007

[36] Michaelis L, Hill ES. The viologen indicators. The Journal of General Physiology. 1933;**16**(6):859-873

[37] Nakamura T, Kawasaki N, Ogawa H, Tanada S, Kogirima M, Imaki M. Adsorption removal of paraquat and diquat onto activated carbon at different adsorption temperature. Toxicological and Environmental Chemistry. 1999; **70**(3–4):275-280

[38] Akinwolemiwa B, Peng C, Chen GZ. Redox electrolytes in supercapacitors. Journal of the Electrochemical Society. 2015;**162**(5):A5054-A5059

[39] Frackowiak E, Fic K, Meller M, Lota G. Electrochemistry serving people and nature: High-energy ecocapacitors based on redox-active electrolytes. ChemSusChem. 2012;**5**(7): 1181-1185

[40] Li Q, Li K, Sun C, Li Y. An investigation of Cu^{2+} and Fe^{2+} ions as active materials for electrochemical redox supercapacitors. Journal of Electroanalytical Chemistry (Lausanne, Switzerland). 2007;**611**(1–2):43-50

[41] Gorska B, Bujewska P, Fic K. Thiocyanates as attractive redox-active electrolytes for high-energy and environmentally-friendly electrochemical capacitors. Physical Chemistry Chemical Physics. 2017; **19**(11):7923-7935

[42] Zhang Y, Zu L, Lian H, Hu Z, Jiang Y, Liu Y, et al. An ultrahigh performance supercapacitors based on simultaneous redox in both electrode and electrolyte. Journal of Alloys and Compounds. 2017;**694**:136-144

[43] Gao Z, Zhang L, Chang J, Wang Z, Wu D, Xu F, et al. Catalytic electrode-redox electrolyte supercapacitor system with enhanced capacitive performance. Chemical Engineering Journal. 2018;**335**: 590-599

[44] Senthilkumar ST, Selvan RK, Lee YS, Melo JS. Electric double layer capacitor and its improved specific capacitance using redox additive electrolyte. Journal of Materials Chemistry A. 2013;**1**(4):1086-1095

[45] Yamazaki S, Ito T, Murakumo Y, Naitou M, Shimooka T, Yamagata M, et al. Hybrid capacitors utilizing halogen-based redox reactions at interface between carbon positive electrode and aqueous electrolytes. Journal of Power Sources. 2016;**326**: 580-586

[46] Bujewska P, Gorska B, Fic K. Redox activity of selenocyanate anion in electrochemical capacitor application. Synthetic Metals. 2019;**253**:62-72

[47] Iamprasertkun P, Ejigu A, Dryfe RAW. Understanding the electrochemistry of "water-in-salt" electrolytes: Basal plane highly ordered pyrolytic graphite as a model system. Chemical Science (Cambridge). 2020; **11**(27):6978-6989

[48] Lee J, Choudhury S, Weingarth D, Kim D, Presser V. High Performance Hybrid Energy Storage with Potassium Ferricyanide Redox Electrolyte. ACS Appl Mater Interfaces. 2016;**8**(36): 23676-23687

[49] Veerasubramani GK, Krishnamoorthy K, Kim SJ. Improved electrochemical performances of binder-free CoMoO4 nanoplate arrays@Ni foam electrode using redox additive electrolyte. Journal of Power Sources. 2016;**306**:378-386

[50] Lamiel C, Lee YR, Cho MH, Tuma D, Shim J-J. Enhanced electrochemical performance of nickel-cobalt-oxide@reduced graphene oxide// activated carbon asymmetric supercapacitors by the addition of a redox-active electrolyte. Journal of Colloid and Interface Science. 2017;**507**: 300-309

[51] Cha SM, Nagaraju G, Chandra Sekhar S, Yu JS. A facile drop-casting approach to nanostructured copper oxide-painted conductive woven textile as binder-free electrode for improved energy storage performance in redox-additive electrolyte. Journal of Materials Chemistry A, Materials for Energy and Sustainability. 2017;**5**(5):2224-2234

[52] Shanmugavani A, Kaviselvi S, Sankar KV, Selvan RK. Enhanced electrochemical performances of PANI using redox additive of $K_4[Fe(CN)_6]$ in aqueous electrolyte for symmetric supercapacitors. Materials Research Bulletin. 2015;**62**:161-167

[53] Su LH, Zhang XG, Mi CH, Gao B, Liu Y. Improvement of the capacitive performances for Co-Al layered double hydroxide by adding hexacyanoferrate into the electrolyte. Physical Chemistry Chemical Physics: PCCP. 2009;**11**(13): 2195-2202

[54] Roldán S, González Z, Blanco C, Granda M, Menéndez R, Santamaría R. Redox-active electrolyte for carbon nanotube-based electric double layer capacitors. Electrochimica Acta. 2011; **56**(9):3401-3405

[55] Yamamoto T, Kanbara T. Porous and electrically conducting clay-carbon composite as positive electrodes of zinc-oxygen primary cells and zinc-iodine secondary cells. Inorganica Chimica Acta. 1988;**142**(2):191-193

[56] Banik SJ, Akolkar R. Suppressing dendrite growth during zinc electrodeposition by PEG-200 additive. Journal of the Electrochemical Society. 2013;**160**(11):D519-D523

[57] Cheng X-B, Zhao M-Q, Chen C, Pentecost A, Maleski K, Mathis T, et al. Nanodiamonds suppress the growth of lithium dendrites. Nature Communications. 2017;**8**(1):336-339

[58] Li B, Nie Z, Vijayakumar M, Li G, Liu J, Sprenkle V, et al. Ambipolar zinc-polyiodide electrolyte for a high-energy density aqueous redox flow battery. Nature Communications. 2015;**6**(1):6303

[59] Bai P, Li J, Brushett FR, Bazant MZ. Transition of lithium growth mechanisms in liquid electrolytes. Energy & Environmental Science. 2016; **9**(10):3221-3229

[60] Fic K, Morimoto S, Frackowiak E, Ishikawa M. Redox activity of bromides in carbon-based electrochemical capacitors. Batteries & Supercaps. 2020; **3**(10):1080-1090

[61] Senthilkumar ST, Selvan RK, Ponpandian N, Melo JS. Redox additive aqueous polymer gel electrolyte for an electric double layer capacitor. RSC Advances. 2012;**2**(24):8937-8940

[62] Lota G, Frackowiak E. Striking capacitance of carbon/iodide interface. Electrochemistry Communications. 2009;**11**(1):87-90

[63] Yu S, Yang N, Zhuang H, Mandal S, Williams OA, Yang B, et al. Battery-like supercapacitors from diamond networks and water-soluble redox electrolytes. Journal of Materials Chemistry A. 2017; **5**(4):1778-1785

[64] Guin PS, Das S, Mandal PC. Electrochemical reduction of quinones in different media: A review. International Journal of Electrochemistry. 2010;**2011**: 1-22

[65] Gastol D, Walkowiak J, Fic K, Frackowiak E. Enhancement of the carbon electrode capacitance by brominated hydroquinones. Journal of Power Sources. 2016;**326**:587-594

[66] Huskinson B, Marshak MP, Suh C, Er S, Gerhardt MR, Galvin CJ, et al. A metal-free organic-inorganic aqueous flow battery. Nature (London). 2014; **505**(7482):195-198

[67] Nie YF, Wang Q, Chen XY, Zhang ZJ. Nitrogen and oxygen functionalized hollow carbon materials: The capacitive enhancement by simply incorporating novel redox additives into H_2SO_4 electrolyte. Journal of Power Sources. 2016;**320**:140-152

[68] Zhang ZJ, Zhu YQ, Chen XY, Cao Y. Pronounced improvement of supercapacitor capacitance by using redox active electrolyte of p-phenylenediamine. Electrochimica Acta. 2015;**176**:941-948

[69] Chen L, Chen Y, Wu J, Wang J, Bai H, Li L. Electrochemical supercapacitor with polymeric active electrolyte. Journal of Materials Chemistry A, Materials for Energy and Sustainability. 2014;**2**(27):10526-10531

[70] Nie YF, Wang Q, Chen XY, Zhang ZJ. Synergistic effect of novel redox additives of p-nitroaniline and dimethylglyoxime for highly improving the supercapacitor performances. Physical Chemistry Chemical Physics. 2016;**18**(4):2718-2729

[71] Wu J, Yu H, Fan L, Luo G, Lin J, Huang M. A simple and high-effective electrolyte mediated with p-phenylenediamine for supercapacitor. Journal of Materials Chemistry. 2012; **22**(36):1925-1193

[72] Chen L, Bai H, Huang Z, Li L. Mechanism investigation and

suppression of self-discharge in active electrolyte enhanced supercapacitors. Energy & Environmental Science. 2014; 7(5):1750-1759

[73] Lezna RO, Centeno SA. Spectroelectrochemistry of methyl viologen/iodide solutions at mercury film electrodes. Langmuir. 1996;**12**(20): 4905-4908

[74] Andreas HA, Lussier K, Oickle AM. Effect of Fe-contamination on rate of self-discharge in carbon-based aqueous electrochemical capacitors. Journal of Power Sources. 2009;**187**(1):275-283

[75] Mourad E, Coustan L, Lannelongue P, Zigah D, Mehdi A, Vioux A, et al. Biredox ionic liquids with solid-like redox density in the liquid state for high-energy supercapacitors. Nature Materials. 2017;**16**(4):446-453

[76] Xie HJ, Gélinas B, Rochefort D. Redox-active electrolyte supercapacitors using electroactive ionic liquids. Electrochemistry Communications. 2016;**66**:42-45

[77] Dall'Agnese Y, Rozier P, Taberna P-L, Gogotsi Y, Simon P. Capacitance of two-dimensional titanium carbide (MXene) and MXene/carbon nanotube composites in organic electrolytes. Journal of Power Sources. 2016;**306**:510-515

[78] Jäckel N, Weingarth D, Schreiber A, Krüner B, Zeiger M, Tolosa A, et al. Performance evaluation of conductive additives for activated carbon supercapacitors in organic electrolyte. Electrochimica Acta. 2016;**191**:284-298

[79] Salunkhe RR, Young C, Tang J, Takei T, Ide Y, Kobayashi N, et al. A high-performance supercapacitor cell based on ZIF-8-derived nanoporous carbon using an organic electrolyte. Chemical Communications (Cambridge, England). 2016;**52**(26):4764-4767

[80] Yang W, Yang W, Song A, Gao L, Su L, Shao G. Supercapacitance of nitrogen-sulfur-oxygen co-doped 3D hierarchical porous carbon in aqueous and organic electrolyte. Journal of Power Sources. 2017;**359**:556-567

[81] Xie L, Sun G, Su F, Guo X, Kong Q, Li X, et al. Hierarchical porous carbon microtubes derived from willow catkins for supercapacitor applications. Journal of Materials Chemistry A, Materials for Energy and Sustainability. 2016;**4**(5): 1637-1646

[82] Zhao C, Zheng W. A review for aqueous electrochemical supercapacitors. Frontiers in Energy Research. 2015;**3**:23

[83] Yamazaki S, Ito T, Yamagata M, Ishikawa M, editors. Nonaqueous Electrochemical Capacitor Utilizing Electrolytic Redox Reactions of Bromide Species in Ionic Liquid. 2012

[84] Yu H, Wu J, Fan L, Hao S, Lin J, Huang M. An efficient redox-mediated organic electrolyte for high-energy supercapacitor. Journal of Power Sources. 2014;**248**:1123-1126

[85] Gong K, Fang Q, Gu S, Li SFY, Yan Y. Nonaqueous redox-flow batteries: Organic solvents, supporting electrolytes, and redox pairs. Energy & Environmental Science. 2015;**8**(12): 3515-3530

[86] Austen Angell C, Ansari Y, Zhao Z. Ionic liquids: Past, present and future. Faraday Discussions. 2011;**154**:9-27

[87] Mousavi MPS, Wilson BE, Kashefolgheta S, Anderson EL, He S, Bühlmann P, et al. Ionic liquids as electrolytes for electrochemical

double-layer capacitors: Structures that optimize specific energy. ACS Applied Materials & Interfaces. 2016;**8**(5): 3396-3406

[88] Brandt A, Pohlmann S, Varzi A, Balducci A, Passerini S. Ionic liquids in supercapacitors. MRS Bulletin. 2013; **38**(7):554-559

[89] Tsai W-Y, Lin R, Murali S, Li Zhang L, McDonough JK, Ruoff RS, et al. Outstanding performance of activated graphene based supercapacitors in ionic liquid electrolyte from −50 to 80 °C. Nano Energy. 2013;**2**(3):403-411

[90] Dubal DP, Suarez-Guevara J, Tonti D, Enciso E, Gomez-Romero P. A high voltage solid state symmetric supercapacitor based on graphene–polyoxometalate hybrid electrodes with a hydroquinone doped hybrid gel-electrolyte. Journal of Materials Chemistry A, Materials for Energy and Sustainability. 2015;**3**(46):23483-23492

[91] Navalpotro P, Palma J, Anderson M, Marcilla R. High performance hybrid supercapacitors by using para-benzoquinone ionic liquid redox electrolyte. Journal of Power Sources. 2016;**306**:711-717

[92] Sathyamoorthi S, Suryanarayanan V, Velayutham D. Organo-redox shuttle promoted protic ionic liquid electrolyte for supercapacitor. Journal of Power Sources. 2015;**274**:1135-1139

[93] Armelin E, Pérez-Madrigal MM, Alemán C, Díaz DD. Current status and challenges of biohydrogels for applications as supercapacitors and secondary batteries. Journal of Materials Chemistry A, Materials for Energy and Sustainability. 2016;**4**(23):8952-8968

[94] Yamagata M, Soeda K, Ikebe S, Yamazaki S, Ishikawa M. Chitosan-based gel electrolyte containing an ionic liquid for high-performance nonaqueous supercapacitors. Electrochimica Acta. 2013;**100**:275-280

[95] Menzel J, Frąckowiak E, Fic K. Agar-based aqueous electrolytes for electrochemical capacitors with reduced self-discharge. Electrochimica Acta. 2020;**332**:135435

[96] Niu J, Conway BE, Pell WG. Comparative studies of self-discharge by potential decay and float-current measurements at C double-layer capacitor and battery electrodes. Journal of Power Sources. 2004;**135**(1–2):332-343

[97] Alipoori S, Mazinani S, Aboutalebi SH, Sharif F. Review of PVA-based gel polymer electrolytes in flexible solid-state supercapacitors: Opportunities and challenges. Journal of Energy Storage. 2020;**27**:101072

[98] Ma G, Dong M, Sun K, Feng E, Peng H, Lei Z. A redox mediator doped gel polymer as an electrolyte and separator for a high performance solid state supercapacitor. Journal of Materials Chemistry A, Materials for Energy and Sustainability. 2015;**3**(7):435-441

[99] Tu QM, Fan LQ, Pan F, Huang JL, Gu Y, Lin JM, et al. Design of a novel redox-active gel polymer electrolyte with a dual-role ionic liquid for flexible supercapacitors. Electrochimica Acta. 2018;**268**:562-568

[100] Fan L-Q, Tu Q-M, Geng C-L, Huang J-L, Gu Y, Lin J-M, et al. High energy density and low self-discharge of a quasi-solid-state supercapacitor with carbon nanotubes incorporated redox-active ionic liquid-based gel polymer electrolyte. Electrochimica Acta. 2020;**331**:135425

[101] Feng E, Ma G, Sun K, Yang Q, Peng H, Lei Z. Toughened redox-active

hydrogel as flexible electrolyte and separator applying supercapacitors with superior performance. RSC Advances. 2016;**6**(79):75896-77594

[102] Hashemi M, Rahmanifar MS, El-Kady MF, Noori A, Mousavi MF, Kaner RB. The use of an electrocatalytic redox electrolyte for pushing the energy density boundary of a flexible polyaniline electrode to a new limit. Nano Energy. 2018;**44**:489-498

[103] Wang Y, Chang Z, Qian M, Zhang Z, Lin J, Huang F. Enhanced specific capacitance by a new dual redox-active electrolyte in activated carbon-based supercapacitors. Carbon (New York). 2019;**143**:300-308

[104] Kim D, Lee G, Kim D, Yun J, Lee S-S, Ha JS. High performance flexible double-sided micro-supercapacitors with an organic gel electrolyte containing a redox-active additive. Nanoscale. 2016;**8**(34): 15611-15620

Chapter 3

Redox Transitions in Pseudocapacitor Materials: Criteria and Ruling Factors

Sergey N. Pronkin, Nina Yu. Shokina and Cuong Pham-Huu

Abstract

Pseudocapacitance is a phenomenon of charge storage involving redox transitions at the electrode/electrolyte interface. As the result of an electrode potential modulation, one or few components of the electrode and/or electrolyte change its/their oxidation states. The redox reaction may be confined to the interface or propagate into the bulk of the electrode material, thus significantly increasing the charge (and energy) capacitance of the material. The rate and the reversibility of the interfacial redox reaction are the key factors determining the efficiency of charge storage due to pseudocapacitance phenomena. The influence of the characteristics of the interfacial redox reaction on the efficiency of charge storage in pseudocapacitive materials is considered in the current chapter. In particular, the similarities and the differences between the charge storage in batteries and pseudocapacitors are discussed. The analysis of the pseudocapacitive behavior of electrode material using the impedance spectroscopy is presented.

Keywords: pseudocapacitance, ion insertion, interfacial kinetics, impedance, staircase model

1. Introduction

The transition toward sustainable energy grids, based on renewable and non-critical resources, relies strongly on the technologies of efficient and reversible energy storage. Electrochemical energy storage devices (EESD) are among the most promising and versatile contemporary devices [1, 2]. Batteries, supercapacitors, and hybrid cells are the main classes of EESD, providing a wide range of energy and power densities. The energy densities of modern batteries are still inferior comparing to traditional (e.g., diesel) and modern (e.g., liquid hydrogen) fuels. However, high storage efficiency and zero operation emission make them indispensible elements of sustainable energy grids (**Table 1**).

Supercapacitors have ca. 10 times lower energy density comparing to batteries, but may operate at much higher power and have high capacitance retention with cycling. These differences are related to different mechanisms of charge storage in batteries and supercapacitors. Namely, in batteries the interfacial electrochemical reaction

Storage device	Energy density (W.h/kg)	Power density (W/kg)	Charge/discharge time	Cycles lifetime
Electrostatic capacitors	0.01–0.1	$>10^5$	ms	∞
Supercapacitors	10–100	1000–10,000	From s to h	$>10^4$
Li batteries	100–300	10–300	Hours	$\approx 10^3$
Diesel fuel	12,667	700[a]	—	1
Liquid hydrogen	39,405	1500–2000[b]	—	≈ 1000 h

[a]*In standard V8 diesel engine.*
[b]*In PEMFC [3].*

Table 1.
Energy and power density of various energy storage devices.

results in the transformation of the bulk of electrode materials. For example, in Li metal batteries, these reactions are the dissolution of Li to Li^+ at negative electrode and the intercalation of Li^+ to positive host electrode (e.g., graphite) [4]. While the maximal energy storage of batteries is primarily determined by the intercalation capacitance of host material, the power of energy input/output is limited by the rates of interfacial charge transfer and ion mass transport in solid-state bulk. Due to these factors, the energy capacitance of electrode materials in batteries strongly depends on the rate of charging.

On the other hand, in electrostatic capacitors, the charge Q is stored on two electrode plates separated by an insulator:

$$C = \frac{dQ}{dE} = \frac{\varepsilon\varepsilon_0}{d}A, \tag{1}$$

where $\varepsilon_0 = 8.854 \cdot 10^{-12}$ F.m^{-2} is the vacuum permittivity, ε is the dielectric constant of the insulator, d is the distance between electrodes, and A is the surface area of the electrode.

Contrary to batteries, the capacitance of a capacitor does not depend on the rate of charging, providing that the charging time is longer than the circuit time constant $\tau = R_{ESR} \cdot C$, where R_{ESR} is the equivalent serial resistance of the circuit (**Figure 1A**).

In electrochemical capacitors, the charge is stored at or in the vicinity of the electrode/electrolyte interface (**Figure 1B**). Comparing to electrostatic capacitors, electrochemical capacitors have much shorter d (order of nm) and higher specific surface area (SSA) A per mass unit. Thus, electrochemical capacitor possesses, in general, the charge capacitance ca. 4 orders of magnitude higher than electrostatic capacitors.

From the practical point of view, it is useful to define various types of specific capacitance, as listed below.

- Surface specific capacitance $C_s = C/A$. According to (1), C_s depends only on the properties of the interface (d and ε).

- Mass specific (gravimetric) capacitance $C_g = C/m$. As a characteristic of supercapacitors, C_g is calculated per total mass of the device. However, in material science C_g is most often calculated per mass m of dry electrode material.

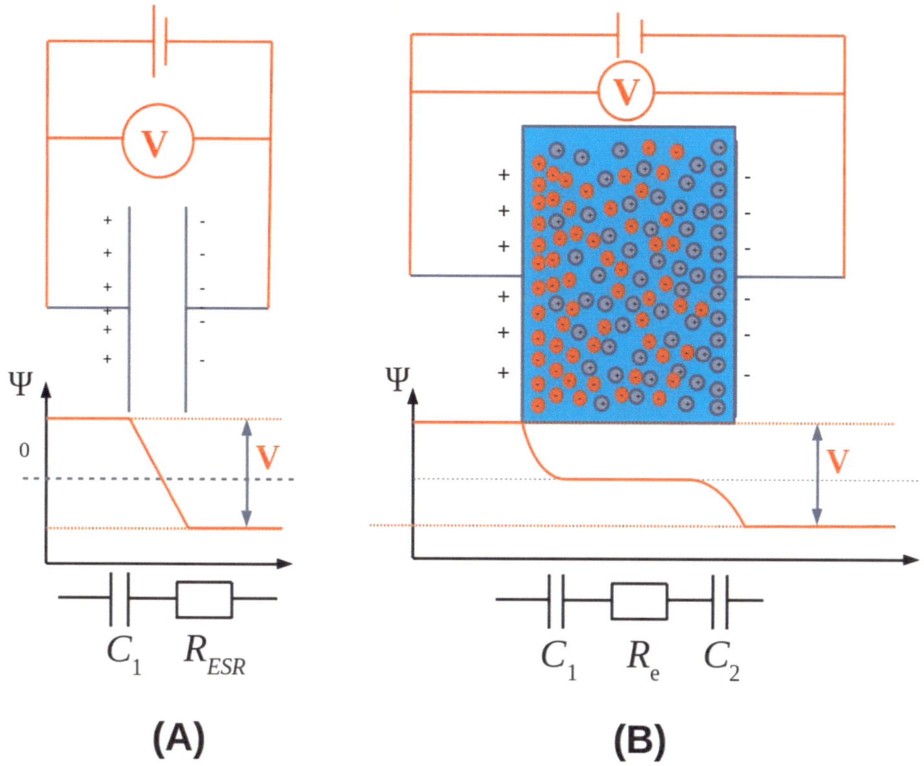

Figure 1.
The scheme of distributions of the electrical potential between the electrodes of a solid-state capacitor (A) and an electrical double layer capacitor (B).

In this case, $C_g = C_s \cdot SSA$. C_g is a useful parameter for comparison of different electrode materials, because C_g depends on both interfacial charge storage efficiency (characterized by C_s) and material structure (characterized by SSA). However, this comparison can be misleading due to the utilization of dry material mass with pores filled by air. The efficient electrode material has an open porous structure with pores filled by electrolyte. It results in a significant increase in the mass of the electrode material. Thus, the true value of C_s is much lower than the value calculated with the mass of dry material.

- Volume specific capacitance $C_V = C/V$. Similarly to C_g, in material science the value of C_V is most often calculated using the volume unit of the electrode material V: $C_V = C_s \cdot SSA \cdot \rho$, where ρ is the density of a porous material. Contrary to C_g, C_V has the same value for dry material and material filled with an electrolyte. This is one of the reasons why C_V is a more relevant characteristics of the charge capacitance of the material than C_g [5].

Depending on the main mechanism of charge storage, one distinguishes between 2 types of supercapacitors: electrochemical double layer capacitors (EDLC) and pseudocapacitors.

As the name suggests, in EDLC main mechanism of charge storage is the formation of an electrical double layer at the electrode/electrolyte interface. In pseudocapacitors, the predominant mechanism of charge storage is the pseudocapacitance phenomena. This phenomenon involves reversible interfacial electrochemical reaction and intercalation of a solvent component into the bulk of the electrode material. From the chemical point of view, the intercalation reactions in batteries and pseudocapacitors are similar. The important difference is that the rate of intercalation in supercapacitors, as opposed to batteries, is not limited by solid-state ion transport, but limited mostly by material intercalation capacitance and interfacial charge transfer rate.

In the current chapter, the influence of the rate of interfacial charge transfer on the electrochemical performance of electrode materials in supercapacitors is considered. The particular features of EDLC and pseudocapacitor performances are described. The focus of the current work is on the electrode materials of pseudocapacitors, because the charge transfer rate strongly influences the rate of pseudocapacitance phenomena. The influence of pseudocapacitance on the performance of EDLC is briefly discussed. The influence of interfacial charge transfer rate on the material capacitance of pseudocapacitors electrodes is first considered for a model flat electrode/electrolyte interface. Then, the dependence of the capacitance of porous electrode materials on interfacial charge transfer rate is analyzed. To evaluate the role of interfacial charge transfer, the models of interfacial impedance in the presence of pseudocapacitance reaction are considered.

2. The influence of the interfacial charge transfer rate on the performance of supercapacitors

2.1 Electrical double-layer capacitors

In EDLC the predominant mechanism of charge storage is charging of interfacial electrical double layer. This layer consists of adsorbed ions of electrolyte and solvent dipole molecules. The change of electrode potential results in changes of the ratio of adsorbed cations and anions at the interface and re-orientation of solvent dipoles. These phenomena result in charge accumulation at the interface with potential change, similarly to the charging of a capacitor when voltage is applied. The detailed models of the structure of the electrical double layer were developed in the last century [6]. As a very simple approximation, electrical double layer can be represented as a capacitor with one plate being an electrode and another plate being adsorbed ions and solvent molecules (**Figure 1B**). Under this approximation, the charge capacitance of the double layer C_{dl} is given by (1).

Considering that for aqueous solutions $\varepsilon = 80$ and the thickness of electrical double layer is $d \approx 1$ nm, the surface specific charge capacitance C_s of electrode/electrolyte interface can be estimated as $C_s \approx 2 \cdot 10^{-5}$ F.cm^{-2} = 20 μF.cm^{-2}. A small thickness d of electrical double layer allows using the materials with highly developed surface as EDLC electrodes, for example, highly porous carbon materials with $SSA > 2600$ m^2.g^{-1} [7]. On the other hand, carbon/electrolyte interface has lower surface specific capacitance C_s comparing to metal/electrolyte interface, namely below 20 μF.cm^{-2} [8]. For a single-layer sp^2 carbon graphene sheet, the value of 13.5 μF.cm^{-2} is predicted and expected to decrease with the thickness of the carbon stack down to ca. 5 μF.cm^{-2} for the graphite structure. Thus, for carbon materials with highly developed SSA, the specific capacitance C_g related to double layer formation is expected to be in the range of 150–200 F/g.

The modulation of electrode potential results in the change of ion adsorption, solvent dipoles orientation, and, possibly, electrode surface reconstruction [9]. These phenomena lead to the changes of the values of d and ε of (1). Thus, the value of specific interfacial capacitance C_s may depend strongly on electrode potential. This results in the a complex shape of CV curves with well-defined CV peaks and different values of double layer charging current at different potentials.

However, these interfacial phenomena are relatively fast and the total charge associated with them does not depend on the rate of charging (rate of potential sweep). Therefore, the value of charging current is expected to be proportional to the rate of potential sweep for a given potential. This proportionality is one of the criteria that was proposed to distinguish between the electrode materials behaving as supercapacitor or battery electrode materials [10].

The surface-specific charge capacitance can be increased due to the phenomena of pseudocapacitance, which involves a fast and reversible surface electrochemical transformation of an electrode component at a certain potential. The following electrochemical phenomena may occur: oxidation/reduction of surface oxides, partial charge transfer to/from adsorbed electrolyte species (for example, electrochemical adsorption/desorption of atomic hydrogen [11], underpotential deposition of electrolyte metal cations [12]).

For carbon electrodes, the characteristic surface redox transformation is attributed to the presence of quinone-type surface groups on the partially oxidized surface [13, 14]. This reversible surface transition is reflected by the appearance of a pair of CV peaks around the equilibrium potential $E_0 = 0.668$ V (RHE) of quinone-hydroquinone redox couple. The surface density of quinone groups on oxidized carbon surfaces is estimated to be approximately equal to $0.1 - 1.1\%_{at.}$ or $10^{-10} - 10^{-11}$ mol.cm^{-2} [14]. These groups provide additional $1 - 10$ μF.cm^{-2} to surface-specific capacitance. The interfacial capacitance is also increased in the presence of heteroatoms in carbon structure. In particular, an increase in surface charge capacitance due to N-doping of carbon was studied most thoroughly [15–18]. N-doping of carbon materials up to few $\%_{wt.}$ can be achieved, resulting in a significant increase in surface specific capacitance. It allows to reach the mass specific capacitance close or even higher than 300 F/g for N-doped carbon electrode materials [17, 18].

As mentioned above, the predominant mechanism of charge storage in EDLC is a charging of an interfacial electrical double layer by ion adsorption and solvent molecule orientation, even though the surface pseudocapacitance phenomena may provide additional contributions. In contrast, in pseudocapacitors, the contribution of pseudocapacitance is predominant due to the propagation of the surface pseudocapacitance reaction further to the bulk of electrode material.

2.2 Pseudocapacitors

Even in the case of carbon materials, the pseudocapacitance reaction may propagate from interface further to the bulk of electrode material. One interesting example of this phenomenon is a hydrogen intercalation into a graphitic carbon structure. This phenomenon has been studied in details on activated carbon CH900–20 ACC (Japan) with SSA = 1520 m^2.g^{-1} [19, 20]. Few hours of galvanostatic cathodic polarization of CH900–20 resulted in a formation of an intercalation compound with stoichiometry C_6H:

$$C_{6(s)} + H^+_{(l)} + e^- \rightleftharpoons C_6H_{(s)} \tag{2}$$

The maximal charge capacitance provided by the reaction (2) is 943 F/g, and the total mass specific capacitance C_g of CH900–20 in acid electrolytes is equal to 1114.3 F/g. To the knowledge of the authors, this is the highest mass-specific capacitance reported for carbon materials. On the other hand, the estimated diffusion coefficient D of hydrogen in carbon, according to [20] is estimated as $5 \cdot 10^{-17}$ $cm^2.s^{-1}$, and the intercalation is controlled by solid-state ion transport. This makes the material CH900–20 to be impractical for supercapacitors.

The pseudocapacitance is the main mechanism of charge storage in transition metal oxide (TMO) electrode materials commonly used as pseudocapacitor electrodes [21–23]. Similar to the functional groups at the carbon surface considered above, the metal cations at or in the vicinity of the electrode/electrolyte interface can undergo a fast reversible transformation due to a variability of oxidation states of transition metals. In comparison to the carbon, the percentage of metal sites available for the redox transformation is much higher: every metal atom is capable to change its oxidation state and participate in charge storage. This phenomenon provides significantly higher pseudocapacitance values for metal oxides comparing to carbon materials. The values above 900 F/g are routinely observed for RuO_2-based electrode materials, which are among the best performing oxide materials in pseudocapacitors. As for the carbon electrodes, different phenomena can be involved in the pseudocapacitance: the surface redox transformation (involving interfacial charge transfer) and ion intercalation into solid-state electrode. For example, for RuO_2, which is the most studied metal oxide supercapacitor material, the redox transition is described as follows [24]:

$$Ru^{4+}O_2 + \delta H^+ + \delta e^- \rightleftharpoons H_\delta Ru^{(4-\delta)+}O_2 \tag{3}$$

The transformation starts as an electroreduction of RuO_2 surface and then propagates into the bulk of the oxide. According to (3), this process can be also considered as H-intercalation. The transformation degree and the stored charge are characterized by the intercalation parameter δ: $0 \leqslant \delta \leqslant 1$.

In more general case, the reaction (3) can be presented as follows:

$$M_xO_y + \delta Ct^+ + \delta e^- \rightleftharpoons M_xO_yCt_\delta \tag{4}$$

The nature of intercalating cation into the oxide materials from aqueous electrolytes is still debated even for Mn oxides—the second most studied type of oxides for supercapacitor electrodes. Most of transition metal oxides utilized in supercapacitor electrodes are stable only in neutral and alkaline solutions. Thus, the intercalation of cations of electrolyte (K^+, Na^+) was proposed to be a pseudocapacitance reaction (4) [25, 26]. The inclusion of Na and S elements into thin film of MnO_2 and the decrease in Mn oxidation state from +4 to +3 (as the result of its polarization) have been indeed confirmed by ex situ XPS study [27]. However, the inclusion of these elements was mostly confined to the sub-surface atomic layers of oxide, while the high value of charge storage capacitance suggests the propagation of pseudocapacitance reaction into the bulk of oxide [27, 28]. Moreover, the pseudocapacitance reaction for Mn oxides appears to be pH-dependent: in general, higher charge capacitance is observed at higher pH [29, 30]. For Mn_3O_4/C electrode material, the potential of pseudocapacitance reaction (4) was found to shift by 59 mV/pH (**Figure 2**).

These facts are more consistent with the nature of pseudocapacitance reaction of Mn oxides in aqueous solutions as expressed by the following equation:

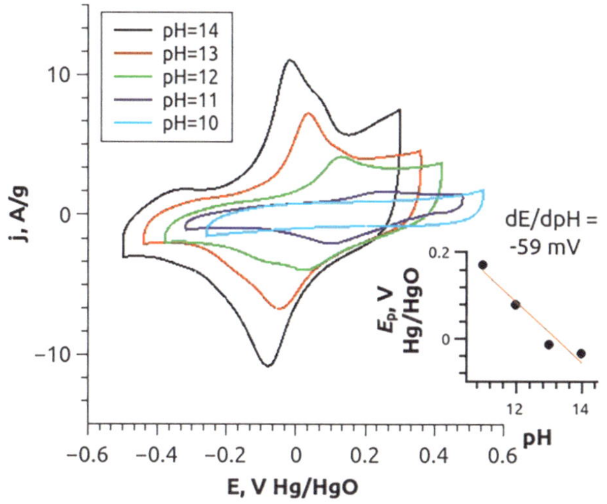

Figure 2.
CV curves of Mn_3O_4/C (34%$_{wt.}$) oxide electrode in 1 M (Na_2SO_4 + NaOH) electrolytes with various pH, measured at 20 mV/s [30].

$$Mn^{4+}O_{x,(s)} + \delta H_2O + \delta e^- \rightleftharpoons Mn^{(4-\delta)+}O_{x-\delta}OH_{\delta,(s)} + \delta OH^-_{(l)} \quad (5)$$

The total specific amount of charge (in C/g) stored due to pseudocapacitance Q_{pc} is as:

$$Q_{pc} = \frac{x \cdot \delta \cdot F}{M}, \quad (6)$$

where x is the stoichiometric coefficient of metal cation in M_xO_y oxide, M is molecular weight of the oxide (in g/mol). For example, for RuO_2, Q_{cp} = 725.4 C/g, according to (6), which is close to the experimental values of specific charge density $Q_C = 614$ C/g reported for amorphous RuO_2 [24]. This indicates that $\delta \approx 0.85$ for RuO_2, i.e., the most of Ru cations in the bulk of the oxide are involved in the charge storage.

2.3 Pseudocapacitors versus batteries

The propagation of reaction (4) into the bulk of the oxide results in the cation Ct^+ intercalation into the oxide M_xO_y. Cation intercalation is also a charge storage mechanism in various types of batteries: in particular, the Li^+ cations are intercalated into the graphite cathodes in the conventional Li batteries [3]. The apparent similarities between these intercalation processes sparkled the discussion about the criteria to distinguish between the materials behavior as battery or as supercapacitor electrodes [10, 31].

As discussed in Section 1, comparing to batteries, the pseudocapacitors provide smaller energy density storage, but operate with higher output power and better reversibility, i.e., retention of capacitance in a larger number of cycles. It suggests that the intercalation reaction is faster and more reversible in supercapacitors. The criteria of reversibility of intercalation reaction are considered below.

The rate of the reversible interfacial redox process of Ct^+ intercalation (4) is expressed by the equation [32], similar to the Buttler-Volmer equation combined with the Frumkin isotherm:

$$i = i_0\left((1-\delta)e^{-(1-\alpha)g\delta}e^{\frac{(1-\alpha)F(E-E_0)}{RT}} - \delta e^{\alpha g\delta}e^{-\frac{\alpha F(E-E_0)}{RT}}\right). \quad (7)$$

Here i_0 is the exchange current density of reaction (4), α is the symmetry factor ($\alpha = 0.5$ for reversible intercalation). The constant g of the Frumkin isotherm characterizes the lateral interaction between the intercalated ions: $g < 0$ for attractive and $g > 0$ for repulsive interaction (most commonly observed for intercalation cations). For example, $g = -4.2$ was found for intercalation of Li^+ into Li_xCoO_2 [32].

The (7) assumes that the cations Ct^+ in the electrolyte and in the solid oxide phase are in quasi-equilibrium, which depends only on the interfacial potential E according to the Frumkin isotherm:

$$\frac{\delta}{1-\delta} = e^{\frac{F(E-E_0)}{RT}}e^{-g(\delta-0.5)}. \quad (8)$$

Eqs. (7) and (8) assume that the distribution of δ within the bulk of host material is not influencing the interfacial reaction, i.e., that the rate of (4) is determined by its kinetics and not by the mass transport of Ct^+ in the solid phase. For example, for Li intercalation into $LiCoO_2$, this model is valid at a potential sweep rate 10–50 $\mu V/s$. At faster charging rates, the rate of Li^+ intercalation is determined by its diffusion in the host material and the Eqs. (7) and (8) are no longer valid.

Figure 3 illustrates the intercalation process of Ct^+ into the oxide particle and the formation of a gradient of Ct^+ concentration (or δ) within the particle. If $\delta_L \gg d$, where δ_L is a characteristic length of concentration gradient and d is the particle size,

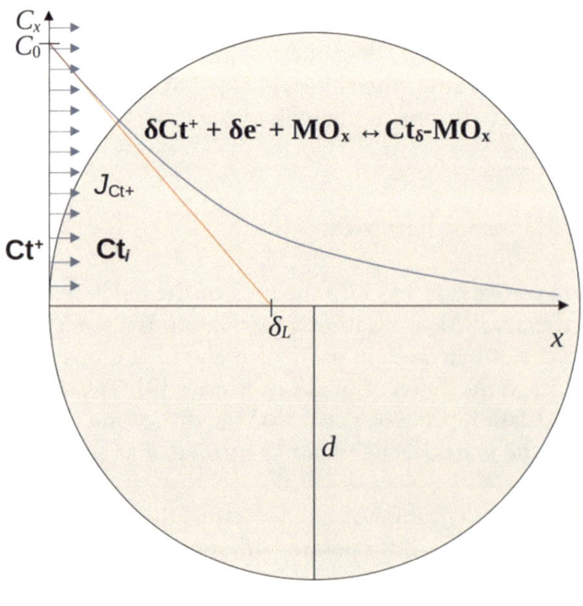

Figure 3.
The intercalation of Ct^+ into the oxide particle.

then the gradient of δ has no influence on the rate of interfacial reaction, the Eqs. (7) and (8) are valid, and the reaction (4) can be considered as reversible.

Assuming that the material of electrode has sufficiently high electronic conductivity, the presence of a gradient of potential inside the solid-state material can be neglected. Thus, the diffusion is a predominant mode of propagation of reaction (4) into the bulk of solid state. The solution of the first Fick's law for planar diffusion with boundary conditions $C_{Ct^+,x=0} = C_0, C_{Ct^+,x=\infty} = 0$ shows the relation between the current I of Ct^+ intercalation reaction (4) and the length of the concentration gradient δ_L:

$$I = AFC_0 \frac{D_{Ct^+}}{\delta_L}. \tag{9}$$

Here A is the surface area of the interface, D_{Ct^+} is the solid-state diffusion coefficient of Ct^+. For an efficient intercalation host material, maximal value of δ is equal to 1 and C_{Ct^+} can be approximated as a concentration of metal cations at the interface, i.e. $C_{Ct^+} = C_0 \cdot \delta$. For example, for MnO_2, $C_0 \approx \rho/M = 0.058$ mol.cm^{-3}.

From the practial point of view, it is useful to relate the charging rate I (in A) with mass specific current density i_m (in A/g) by considering the geometry of a spherical particle with the diameter d:

$$i_m = \frac{6IA}{\rho d}, \tag{10}$$

where ρ is particles density (in g.cm^{-3}). Combining (9) and (10), the equation for the dependence of δ_L on i_m is obtained:

$$\delta_L = \frac{6FC_0\delta D_{Ct^+}}{i_m \rho d}. \tag{11}$$

Eq. (11) provides the criterion of reversibility of reaction (4). If $\delta_L \gg d$, then the concentration of intercalated cation within the oxide is nearly constant. In this case, the value of intercalation factor δ depends only on interfacial conditions, namely potential E, and the Eqs. (7) and (8) are valid. On the other hand, if $\delta_L \approx d$, then the value of intercalation factor δ depends on the distance from interface and is changing with time at constant E.

The strongest incertitude in Eq. (11) is related to the values of solid-state diffusion coefficient D_{Ct^+} at ambient temperature. For Li batteries electrode materials, the values of D_{Li^+} are reported mostly for non-aqueous electrolytes. The values as high as $D_{Li^+/C} = 10^{-9} - 10^{-10}$ cm^2.s^{-1} were determined for Li^+ intercalation into graphite electrode using impedance spectroscopy [33]. However, the values of Li^+ solid-state diffusion coefficients are most commonly found in the range $10^{-13} - 10^{-15}$ cm^2.s^{-1} using the modeling of galvanostatic charging curves [34–36]. The advantages and limitations of this approach are thoroughly discussed in [37]. The thorough review of ambient temperature ionic conductivity in Li batteries electrode material is given in [38].

The diffusion coefficients D_{Ct^+} for cation intercalations from aqueous electrolytes are seldom measured. Indirectly, the values of D_{Ct^+} can be estimated from the ionic conductivity values of oxide σ_i using the following Equation [38]:

$$\sigma_i = \frac{F^2 C_{Ct^+} D_{Ct^+}}{RT}. \tag{12}$$

For example, for MnO_2 oxides, σ_i was reported in the range 0.001–0.02 $Ohm^{-1}.cm^{-1}$, depending on their crystallographic structure [39]. Thus, the diffusion coefficient values can be estimated as $D_{Ct^+} \approx 10^{-7} - 10^{-9}$ $cm^2.s^{-1}$ for a cation insertion into MnO_2 from aqueous electrolytes. In general, the rate of diffusion in supercapacitor electrode materials is expected to be few orders of magnitude larger comparing to Li^+ diffusion in Li batteries electrode materials. Using the Eqs. (11) and (12), one may estimate δ_L value for 100 nm particles of a material with $\sigma_i = 0.001$ $Ohm^{-1}.cm^{-1}$: $\delta_L \approx 0.03$ cm even at fast charging rates ($i_m = 100$ A/g). Thus, for the oxide materials with relatively high ionic conductivity, $\delta_L \gg d$ even for fast charging rates. It indicates that the intercalation reaction can be considered as reversible and the material acts as a pseudocapacitor electrode.

The equations above provide qualitative criteria to distinguish between the materials behaving as supercapacitor or battery electrodes. To develop quantitative criteria, the modeling of various simultaneous interfacial processes as functions of their rates is needed. The simplest strategies for these models are considered below.

2.4 Impedance of flat interface in the presence of pseudocapacitance reaction

Impedance spectroscopy may provides detailed information of the interfacial phenomena occurring with various rates. In particular, in the presence of pseudocapacitance reaction on the flat electrode/electrolyte interface the following processes have to be taken into account.

- Formation of electrical double layer by reversible adsorption of electrolyte ions. Assuming a small amplitude of potential modulation, the interfacial characteristics (dielectric constant ε, change of ion adsorption with potential $d\Gamma/dE$) can be considered to be constant. The impedance of double layer formation is then expressed as a capacitance impedance:

$$Z_{dl} = -\frac{i}{\omega C_{dl}},$$

where $\omega = 2\pi f$ is the angular frequency (in rad/s), f is the potential modulation frequency (in Hz), C_{dl} is the double layer capacitance, $C_{dl} = C_s \cdot A$, where C_s is the surface specific capacitance (in $F.cm^{-2}$), A is the electrode area (in cm^2). Depending on the electrode material, C_s varies between 5 $\mu F.cm^{-2}$ for carbon electrodes and 20–50 $\mu F.cm^{-2}$ for metal electrodes.

- Interfacial charge transfer in the course of the reaction (4). For small potential modulations around the equilibrium redox potential E_0, the Eq. (7) can be simplified to a linear form. The current i is proportional to the overvoltage $E - E_0$ and the impedance of this process can be approximated by a simple resistance R_{ct}:

$$Z_{ct} = R_{ct} = \frac{RT}{F \cdot i_0}.$$

- Diffusion of charged species into the bulk of the oxide material. Assuming the semi-infinite diffusion conditions ($C_{Ct^+,x=0} = C_0, C_{Ct^+,x=\infty} = 0$), the impedance of this process is given by the warburg impedance:

$$Z_W = \frac{\sigma}{\sqrt{\omega}} - \frac{i\sigma}{\sqrt{\omega}},$$
$$\sigma = \frac{RT}{\sqrt{2D}AF^2C_0}.$$

- Accumulation of charge in the bulk of oxide material. For small potential modulation this process can be considered similar to the accumulation of the charge at a capacitor:

$$Z_P = -\frac{i}{\omega C_p},$$
$$C_p = \frac{C_g \cdot A}{SSA \cdot 10^4}.$$

Here C_p is the capacitance of charge accumulation due to the pseudocapacitance reaction (in F), C_g is the gravimetric capacitance of electrode material (in F/g), *SSA* is the specific surface area of electrode material (in $m^2.g^{-1}$).

For a small potential modulation, the listed processes can be considered to be independent. In this case, the total interfacial impedance can be expressed by the Frumkin-Melik-Gaykazyan (FMG) model [32, 33], depicted as an equivalent circuit in **Figure 4**.

From the practical point of view, it is useful to correlate the frequency of the potential modulation f in the impedance spectra with the typical rates of potential variation during the charge/discharge of supercapacitors. The correlation can be done by using the root mean square rate v_{rms} of potential in the course of sinusoidal modulation:

$$v_{rms} = \frac{2\pi E_0 f}{\sqrt{2}}, \tag{13}$$

where E_0 is the amplitude of potential modulation. One can demonstrate that 1C charging rate of a capacitor with 1 V operation voltage is comparable with the v_{rms} at $f = 6.25$ Hz (for $E_0 = 10$ mV). For 10C charging rate, the equivalent frequency is 62.5 Hz. Thus, one can roughly define the frequency range of interest for supercapacitors as ca. 0.1–10 Hz, and for batteries—below 1 Hz. This attribution is important for an interpretation of impedance modeling results.

Figure 4 shows the results of the modeling of interfacial impedance of a flat electrode in the presence of pseudocapacitance reaction, assuming $C_s = 5\ \mu F.cm^{-2}$, $C_g = 90\ F.g^{-1}$, $SSA = 30\ m^2.g^{-1}$. The maximal surface specific capacitance of this interface is $C_m = 305\ \mu F.cm^{-2}$, consisting of both C_{dl} and C_p contributions. The Bode plots (**Figure 4B, D**) show the calculated values of the surface specific interfacial capacitance $C_{s,c}$ as a function of the modulation frequency and the rates of interfacial charge transfer i_0 and solid-state diffusion D. The values of $C_{s,c}$ are calculated from the calculated values of the interfacial impedance Z:

$$C_{s,c} = -\frac{1}{\omega \cdot \mathrm{Im(Z)} \cdot A}. \tag{14}$$

As discussed above, the diffusion rate in supercapacitor electrode materials is few orders of magnitude larger comparing to Li batteries electrode materials. The

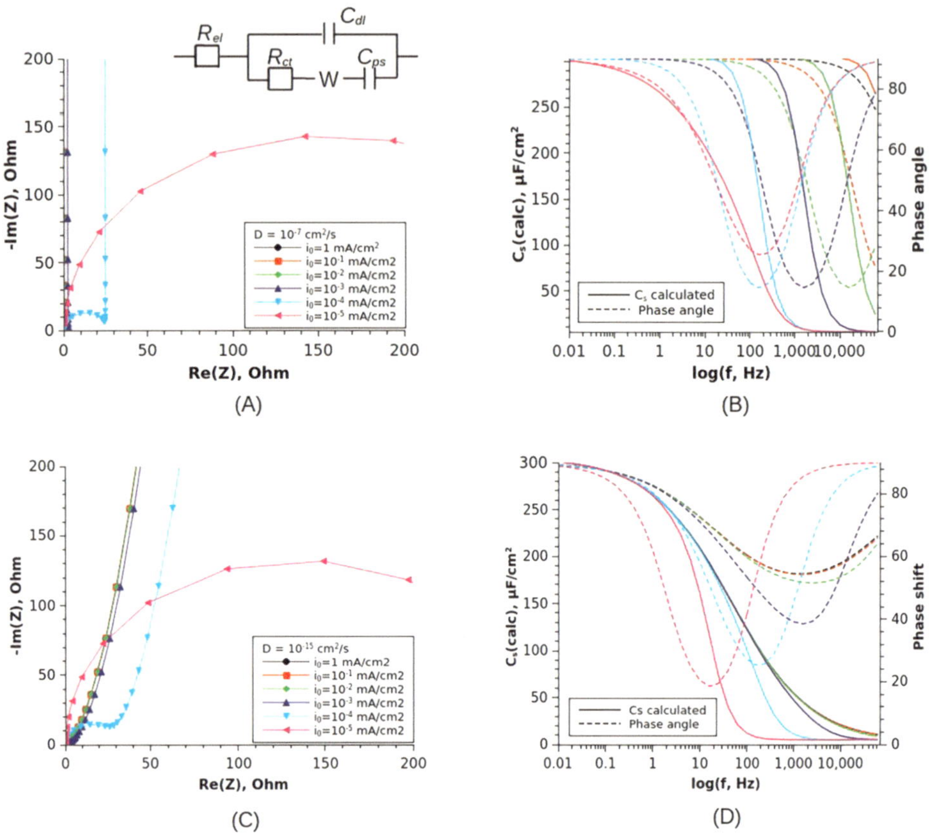

Figure 4.
Nyquist (A,C) and bode (B,D) plots for the Frumkin-Melik-Gaykazuan model of interacial impedance in the presence of pseudocapacitance reaction, calculated for solid-state diffusion coefficients $D = 10^{-7}$ cm^2.s^{-1} *(A,B) and* $D = 10^{-15}$ cm^2.s^{-1} *(C,D) and for various charge transfer rates (depicted in legend).*

influence of the rate of solid-state diffusion on the behavior of the electrode materials can be illustrated by the comparison of the Bode plots calculated by the FMG model with $D = 10^{-7}$ cm^2.s^{-1} for supercapacitors and with $D = 10^{-15}$ cm^2.s^{-1} for batteries electrodes.

In the first case ($D = 10^{-7}$ cm^2.s^{-1}), for not particularly slow charge transfer rate ($i_0 > 10^{-4}$ mA.cm^{-2}) C_s is frequency-independent and close to C_m. This behavior is observed in relatively wide frequency range at low and moderate frequencies ($f < 100$ Hz). This shape of the Bode plots shows that the electrode material, including its bulk, acts as capacitor, while neither rate of charge transfer, nor rate of diffusion, are limiting factors of its charge/discharge. This is the set of conditions under which the material acts as a pseudocapacitor. Under these conditions, the amount δ of intercalated ions is determined by the interfacial equilibrium according to Frumkin isotherm (8), and depends only on the electrode potential. The phase angle of the interfacial impedance is close to $\phi = 90°$ (dotted curves in **Figure 4B,D**). The Nyquist impedance plots under these conditions show straight vertical line with possible slight inclination due to the constant phase element behavior. The latter is often observed experimentally and explained by the capacitance slight dependence on applied frequency due to the surface heterogeneity, both chemical and geometrical [40, 41].

For very slow charge transfer rate ($i_0 \leq 10^{-5}$ mA.cm^{-2}), a clear decrease in C_s with an increase in f is expected even at low frequencies above 0.01 Hz, indicating that the material is not behaving as a pseudocapacitor even despite fast ionic transport in the solid-state.

At slow diffusion in solid state ($D = 10^{-15}$ cm^2.s^{-1}), the surface specific charge capacitance C_s decreases with increase in f even for fast charge transfer rates. In fact, the model curves for exchange rates $i_0 = 1 - 0.01$ mA.cm^{-2} are nearly identical, demonstrating that the rate of charging process is determined by a slow diffusion rate. For lower i_0 the decrease in C_s with increase in f becomes much sharper, as both diffusion and interfacial charge rates become limiting factors. As the amount δ of intercalated ions is limited by these factors, it is no longer in agreement with the Frumkin isotherm and the material cannot be considered as pseudocapacitor electrode material.

The described simple model is valid for flat interface only, because it neglects the effect of the geometry of electrode surface and material deposit. However, in EESD, such as batteries and supercapacitors, porous materials with well-developed interfacial surface area are utilized. To model the impedance of electrodes of supercapacitors, the influence of porous structure has to be taken into account.

2.5 Impedance of porous electrodes for supercapacitors

The interfacial impedance inside a pore is often modeled by the staircase-type equivalent circuit.

It was demonstrated that for the infinitely long cylindrical pore the interfacial impedance Z_p is described by the DeLevi Eq. (15) [42]:

$$Z_p = \sqrt{R|Z|}e^{\frac{i\phi}{2}}, \tag{15}$$

where R is the electrolyte resistance within the pore, Z is the impedance of a pore wall/electrolyte interface, and ϕ is the phase angle of Z. The Eq. (15) predicts that phase angle of the interfacial impedance of the pore is equal to a half of the phase angle of flat interface with the same characteristics. For example, the phase angle of the impedance of the interface with a double layer formation only is 90° for a flat interface, and 45° for the interface inside the infinite pore.

The behavior of the interfacial impedance of a porous system is determined by the penetration depth λ of potential modulation signal into the pore [43]:

$$\lambda = \frac{1}{2}\sqrt{\frac{\sigma d_p}{2\omega C_s}}, \tag{16}$$

here σ is the conductivity of the electrolyte, d_p is the pore diameter. The Eq. (16) suggests that at the frequency below a certain threshold f_0, $\lambda > l_p$, where l_p is the length of the pore, and the whole surface of pore acts as a flat interface. Thus, the transition from infinite pore behavior to flat interface is observed on the impedance curves, that is often detected in the experiments with porous electrodes (for example [44]).

This staircase model allows to correctly fit experimental impedance spectra of porous electrodes in the absence of interfacial faradaic processes and determine the capacitance C_s. Moreover, the average pore length l_p can be estimated from the f_0

value, providing that the average diameter d_p of the pores is known from the other methods [45]. The model can also be adapted to take into account irregular shape of the pores: conical, globular pores etc. [43].

The majority of the studies of the impedance of porous systems simulate the pores as a system of parallel independent pores. The staircase circuit is applied separately to each pore, and total impedance is calculated as a parallel combinations of the pores. The supercapacitors electrode materials, for example the activated carbon, often have branching hierarchical structures: the larger pores are branching into few narrower pores of next generation, which can be branching further. The hierarchical porous structure of these materials makes the application of the staircase model more complicated, because the staircase circuits of pores from different generations are no longer independent.

The model of branching pores was proposed in [46] by introducing 2 generations of pores: the large (μm size) voids between carbon particles and the narrower (sub-μm size) internal pores. The larger pores were branching into n smaller pores with length and diameter scaled by a_l and a_d factors comparing to the parent pore. This model clearly demonstrated the influence of various parameters of geometry of pores on the performance of carbon materials in supercapacitors. An important conclusion of this work was the role of the interplay between the different parameters of the geometry. In particular, as the porosity of the material was a fixed parameter in the model, an increase in branching factor resulted in an increase in the diameter scaling factor, i.e., narrower pores of the next generation. As it follows from (16), the AC-penetration depth λ is shorter for an narrower pore, and, therefore, the utilization of an interface within such pore is less efficient. Thus, it was concluded that the branching of pores is counter-productive for their efficient utilization [46].

The staircase model was recently generalized take into account complexity of the porous structure of carbon electrode materials of supercapacitors [47]. Three generations of pores were considered: short and wide pores ($d_1 = 10 - 30$ nm) of the 1st generation were branching into β_{12} narrow mesopores of the 2nd generation ($d_2 = 3 - 10$ nm). The pores of the 2nd generation formed the main population of pores. These pores could also branch into β_{23} micropores ($d < 1$ nm). The branching was allowed to occur along the whole length of a parent pore. To take into account the contribution of pseudocapacitance, the Z_f interfacial impedance (see **Figure 5**) was

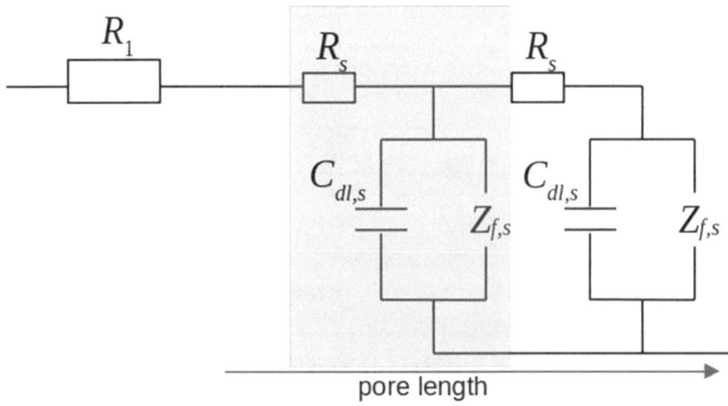

Figure 5.
The staircase equivalent circuit of the impedance model inside a single pore. The highlighted elements form the repeating pattern along the pore.

modeled as a serial combination of charge transfer resistance R_{ct} and pseudocapacitance C_p. To limit the number of model parameters, relatively fast ion transport in the solid phase of the electrode was assumed. Thus, the warburg impedance Z_W was excluded from this model.

The Nyquist plots of the interfacial impedance, modeled by generalized staircase model in the absence of the faradaic impedance, are shown in **Figure 6A**. In agreement with simpler models, the curves show the transition from infinite pore behavior ($\phi = 45°$) at high frequency to flat capacitor behavior ($\phi = 90°$) at frequencies $f > f_0$. In the presence of the fast pseudocapacitance reaction ($i_0 = 0.1$ mA/cm^2) and high pseudocapacitance ($C_p = 100$ μF.cm^{-2}), a characteristic semi-circular behavior is observed at intermediate frequencies (**Figure 6A,C**). The model also shows that the interfacial impedance changes significantly with the geometry of the electrode material for the same characteristics of pseudocapacitance reaction (**Figure 6A,C**).

One may distinguish between two main factors of the influence of the geometry of electrodes on the performance of electrode materials in supercapacitors:" penetration depth" (λ-factor) and" porosity" (ρ-factor). The λ-factor is related to the decrease in the value of λ with an increase in potential modulation frequency f (16). As the

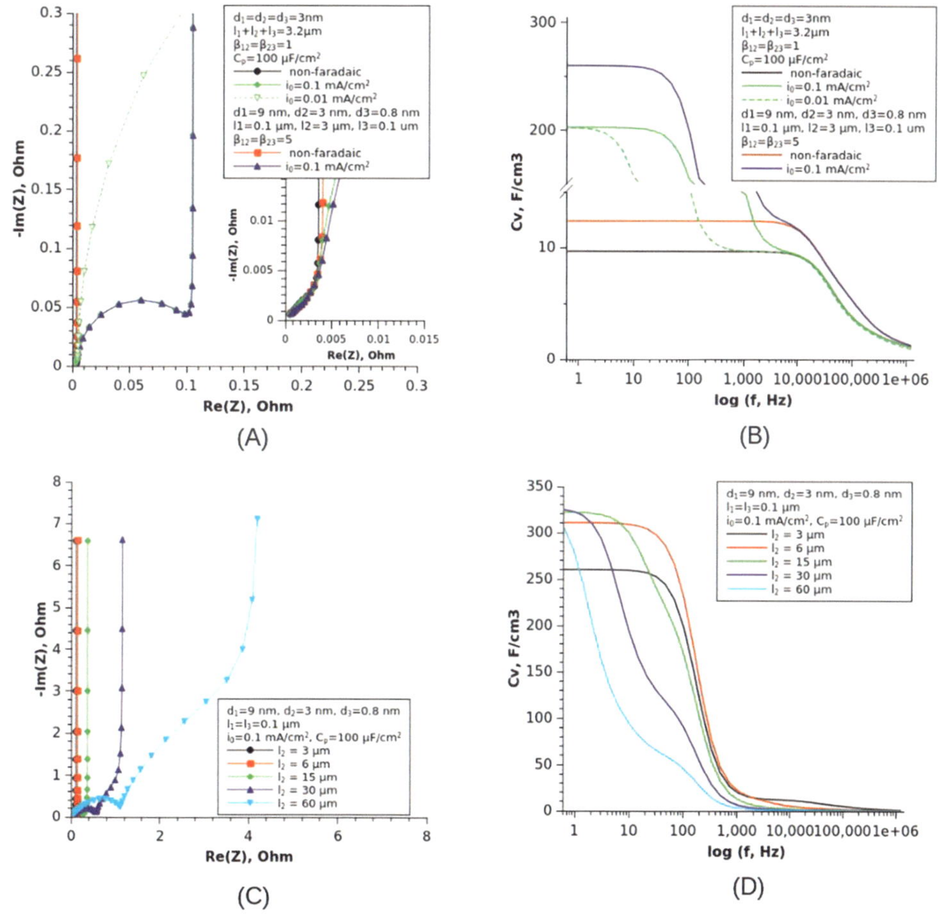

Figure 6.
The Nyquist (A,C) and bode (B,D) plots for porous electrodes calculated by the generalized staircase model with 3 generations of pores. The non-faradaic model excludes Z_f from the calculations.

surface of" active" interface within the pores decreases, the interfacial capacitance C_s is also decreasing with increase in f, and this decrease is more pronounced for smaller and longer pores. On the other hand, smaller pores have lower inner volume and can provide high SSA while having lower porosity and higher density ρ of the material. Thus, these two factors are often found to be counter-active.

As it was mentioned above, the utilization of volume specific capacitance $C_V = C_s \cdot SSA \cdot \rho$ allows accounting for both λ- and ρ- factors. Similar to the previous section 2.4, C_s can be calculated from the model impedance values by (14). Similar to the model of flat interface (**Figure 4B,D**), the Bode plots for porous electrodes, constructed with C_V (**Figure 6B,D**), demonstrate flat segments at low frequencies, corresponding to $\phi = 90°$, or pseudocapacitive behavior. The values of C_V of these segments are determined by ρ-factor: it is higher for smaller and longer pores. Also, the branching of the pores provides the material with higher density (fewer pore are needed for high SSA), and, thus, higher C_V values at low frequencies (**Figure 6B**).

However, comparing to flat interface, the segment of pseudocapacitive behavior is much shorter: the decrease in C_V with increasing frequency is observed at significantly lower frequencies. This decrease is related to λ-factor of geometry influence and it is more significant and observed at lower threshold f_0 for smaller and longer pores (**Figure 6D**).

For fast pseudocapacitance reaction ($i_0 = 0.1$ mA.cm^{-2}) and the typical porous structure of mesoporous carbon (predominant pores with $d \approx 3 - 10$ nm and $l \approx 1 - 3$ μm), the pseudocapacitive behavior is observed up to $f \approx 100$ Hz, i.e., in the region of potential modulation rates relevant for supercapacitors. However, for slower reactions the range of pseudocapacitive behavior is limited to low-frequency ranges only ($f < 1Hz$), showing strong effect of λ-factor of the geometry of pores on the capacitance of electrode material. Also, for longer pores of few tens of μm (e.g., thick electrode deposits), the pseudocapacitive behavior is restricted to slow potential modulation even for very fast pseudocapacitance reaction (**Figure 6D**).

3. Conclusions

The phenomenon of pseudocapacitance allows to increase significantly the charge storage capacitance of the materials by involving the bulk of the electrode material to the charge storage. This phenomenon involves a fast and reversible interfacial charge transfer reaction, followed by an insertion of electrolyte cation into sub-surface and bulk layers of the solid-state electrode. This phenomenon is similar to cation intercalation into the batteries electrodes. However, in the case of supercapacitors, the interfacial reaction must be fast and reversible. In this case, the amount of inserted cation, as characterized by the intercalation factor δ, is in agreement with the Frumkin isotherm (8). In practical terms, this criterion means that the amount of stored charge depends only on the applied potential and does not depend on the rate of charging.

In the case of several simultaneous interfacial phenomena, the electrochemical behavior of the materials can be analyzed by interfacial impedance measurements and modeling. The advantage of the impedance spectroscopy is that, due to a small potential modulation, various interfacial phenomena can be considered to be independent and total impedance can be modeled as a combination of several elements corresponding to different processes. Namely, the impedance of flat electrode/electrolyte interface in the presence of pseudocapacitance phenomena can be analyzed by the FMG model. This analysis shows that in the case of relatively fast ion transport in the solid phase ($D \approx 10^{-7}$ cm^2.s^{-1}) and not particularly slow interfacial charge transfer rates (i.e., $i_0 > 10^{-4}$ mA.cm^{-2}), the pseudocapacitive behavior is expected up to very fast potential modulation ($f \approx 100$ Hz). Under these conditions, the calculated surface specific interfacial capacitance C_s does not depend on the frequency f and the impedance phase angle is close to $\phi = 90°$. This behavior is in clear contrast to the behavior of materials with slow ion transport in solid phase $D \approx 10^{-15}$ cm^2.s^{-1}, in which case a decrease in C_s with increase in potential modulation rate is detected already at low frequencies $f \approx 0.01$ Hz even for fast charge transfer rates ($i_0 = 0.1$ mA.cm^{-2}).

The analysis of the electrochemical behavior of electrode materials for supercapacitors must take into account the complex porous structure of these materials. For this purpose, the generalized staircase model of interfacial impedance for materials with hierarchical branching porous structure can be used. The effects of the geometry of pores on the performance of electrode materials of supercapacitors can be roughly categorized into two groups of factors. First, the presence of pores allows increasing significantly the specific surface area of the material, thus increasing the area of interface available for charge storage. The narrower pores with higher branching factors have lower inner volume, resulting in higher density of electrode material and, correspondingly, higher volume specific capacitance C_V. This group of effects is referred to as ρ-factors. The second group of effects, referred to as λ-factors, is related to the dependence of λ, i.e., penetration depth of potential modulation to the pore, on the potential modulation frequency f. The value of λ is shorter for narrower pores (16); thus ρ- and λ-factors are most often counter-active.

The modeling of the interfacial impedance of porous electrodes shows that the pseudocapacitive behavior (i.e., the values of C_s and C_V being independent on the potential modulation rate f) is confined to lower frequencies f due to λ-factors, comparing to flat interface. In particular, for thick deposits of mesoporous carbon, a decrease in interfacial capacitance with increase in f can be expected even for relatively fast interfacial charge transfer rate $i_0 \approx 0.01$ mA.cm^{-2}.

In general, one may conclude that in the case of flat electrodes of materials with sufficiently high ionic conductivity ($\sigma_i > 0.001$ Ohm^{-1}.cm^{-1}), the pseudocapacitive behavior of electrode materials can be observed even for moderate interfacial charge rates ($i_0 \geqslant 10^{-4}$ mA.cm^{-2}). However, the materials with high *SSA* and developed porous structure are commonly employed in supercapacitors, which imposes stronger limitations on the rate of interfacial charge transfer. Thus, the demonstration of the pseudocapacitive behavior of a material with flat geometry of the interface is not sufficient to suggest its application in supercapacitors. The pseudocapacitive behavior has to be experimentally demonstrated and/or numerically simulated for the given material with porous structure.

Acknowledgements

Sergey Pronkin is indebted to ANR for the financial support of the project INFINE (ANR-21-CE08-0025, 2021-2024).

Conflict of interest

Authors declare no conflict of interests.

Abbreviations

EESD	Electrochemical energy storage device
CV	Cyclic voltammetry
SSA	Specific Surface Area
RHE	Reversible hydrogen electrode
TMO	Transition metal oxides
XPS	X-ray photoelectron spectroscopy
MFG	Frumkin-Melik-Gaykazyan (impedance model)

Author details

Sergey N. Pronkin[1*], Nina Yu. Shokina[2] and Cuong Pham-Huu[1]

1 Institute of Chemistry and Processes for Energy, Environment, and Health (ICPEES UMR-7515 CNRS-Unistra), Strasbourg, France

2 University of Freiburg, Medical Center, Radiology Clinics, Freiburg, Germany

*Address all correspondence to: sergey.pronkin@unistra.fr

References

[1] Grey CP, Tarascon JM. Sustainability and in situ monitoring in battery development. Nature Materials. 2016; **16**(1):45-56. DOI: 10.1038/nmat4777

[2] Dühnen S, Betz J, Kolek M, Schmuch R, Winter M, Placke T. Toward green battery cells: Perspective on materials and technologies. Small Methods. 2020;**4**(7):2000039. DOI: 10.1002/smtd.202000039

[3] Mock P, Schmid SA. Fuel cells for automotive powertrains — A techno-economic assessment. Journal of Power Sources. 2015;**190**(2009):133-140. DOI: 10.1016/j.jpowsour.2008.10.123

[4] Tarascon JM, Armand M. Issues and challenges facing rechargeable lithium batteries. Nature. 2001;**414**(6861): 359-367. DOI: 10.1038/35104644

[5] Gogotsi Y, Simon P. True performance metrics in electrochemical energy sources. Science. 2011;**334** (January):917-918. DOI: 10.1126/science.1213003

[6] Parsons R. Electrical double layer – Recent experimental and theoretical developments. Chemical Reviews. 1990; **90**:813-826

[7] Zhang LL, Zhou R, Zhao XS. Carbon-based materials as supercapacitor electrodes. Journal of Materials Chemistry. 2009;**38**(29):2520-2531. DOI: 10.1039/c000417k

[8] Ji H, Zhao X, Qiao Z, Jung J, Zhu Y, Lu Y, et al. Capacitance of carbon-based electrical double-layer capacitors. Nature Communications. 2014;**5**(3317):1-7. DOI: 10.1038/ncomms4317

[9] Pajkossy T, Wandlowski T, Kolb DM. Impedance aspects of anion adsorption on gold single crystal electrodes. Journal of Electroanalytical Chemistry. 1996; **414**(2):209-220. DOI: 10.1016/0022-0728(96)04700-6

[10] Simon P, Gogotsi Y, Dunn B. Where do batteries end and supercapacitors begin? Science. 2014;**343**(6176): 1210-1211. DOI: 10.1126/science.1249625

[11] Jerkiewicz G. Electrochemical hydrogen adsorption and absorption. Part 1: Under-potential deposition of hydrogen. Electrocatalysis. 2010;**1**(4): 179-199. DOI: 10.1007/s12678-010-0022-1

[12] Herrero E, Buller LJ, Abruña HD. Underpotential deposition at single crystal surfaces of Au, Pt, Ag and other materials. Chemical Reviews. 2001; **101**(7):1897-1930. Available from: http://www.ncbi.nlm.nih.gov/pubmed/11710235

[13] Noked M, Soffer A, Arubach D. The electrochemistry of activated carbonaceous materials: Past, present, and future. Journal of Solid State Electrochemistry. 2011;**15**(7–8):1563-1578. DOI: 10.1007/s10008-011-1411-y

[14] Kinoshita K, Bett JAS. Potentiodynamic analysis of surface oxides on carbon blacks. Carbon. 1973; **11**(4):403-411. DOI: 10.1016/0008-6223(73)90080-8

[15] Borenstein A, Hanna O, Attias R, Luski S, Brousse T, Aurbach D. Carbon-based composite materials for supercapacitor electrodes: A review. Journal of Materials Chemistry A. 2017; **5**(25):12653-12672. DOI: 10.1039/c7ta00863e

[16] Mostazo-López MJ, Ruiz-Rosas R, Morallón E, Cazorla-Amorós D.

Generation of nitrogen functionalities on activated carbons by amidation reactions and Hofmann rearrangement: Chemical and electrochemical characterization. In: Carbon. Vol. 91. Amsterdam, Netherlands: Elsevier; 2015. pp. 252-265. DOI: 10.1016/j.carbon.2015.04.089

[17] Frackowiak E, Lota G, Machnikowski J, Vix-Guterl C, Béguin F. Optimisation of supercapacitors using carbons with controlled nanotexture and nitrogen content. Electrochimica Acta. 2006;**51**(11):2209-2214. DOI: 10.1016/j.electacta.2005.04.080

[18] Wang R, Lang J, Yan X. Effect of surface area and heteroatom of porous carbon materials on electrochemical capacitance in aqueous and organic electrolytes. Science China Chemistry. 2014;**57**(11):1570-1578. DOI: 10.1007/s11426-014-5123-x

[19] Volfkovich YM, Bograchev DA, Mikhalin AA, Bagotsky VS. Supercapacitor carbon electrodes with high capacitance. Journal of Solid State Electrochemistry. 2014;**18**(5):1351-1363. DOI: 10.1007/s10008-013-2271-4

[20] Volfkovich YM, Bograchev DA, Rychagov AY, Sosenkin VE, Chaika MY. Supercapacitors with carbon electrodes. Energy efficiency: Modeling and experimental verification. Journal of Solid State Electrochemistry. 2015;**19**(9): 2771-2779. DOI: 10.1007/s10008-015-2804-0

[21] Simon P, Gogotsi Y. Materials for electrochemical capacitors. Nature Materials. 2008;**7**(11):845-854. DOI: 10.1038/nmat2297

[22] Wang G, Zhang L, Zhang J. A review of electrode materials for electrochemical supercapacitors. Chemical Society Reviews. 2012;**41**(2): 797-828. DOI: 10.1039/c1cs15060j

[23] González A, Goikolea E, Barrena JA, Mysyk R. Review on supercapacitors: Technologies and materials. Renewable and Sustainable Energy Reviews. 2016;**58**: 1189-1206. DOI: 10.1016/j.rser.2015.12.249

[24] Zheng JP. The limitations of energy density for electrochemical capacitors. Journal of the Electrochemical Society. 1997;**144**(6):2026. DOI: 10.1149/1.1837738

[25] Wei W, Cui X, Chen W, Ivey DG. Manganese oxide-based materials as electrochemical supercapacitor electrodes. Chemical Society Reviews. 2011;**40**(3):1697-1721. DOI: 10.1039/C0CS00127A

[26] Tsumura T, Tsumori K, Shimizu G, Toyoda M. Electrochemical properties of spinel-type manganese oxide/porous carbon nanocomposite powders in 1 M KOH aqueous solution. Journal of Physics and Chemistry of Solids. 2012;**73**(2): 237-244. DOI: 10.1016/j.jpcs.2011.10.036

[27] Toupin M, Brousse T, Bélanger D. Charge storage mechanism of MnO_2 electrode used in aqueous electrochemical capacitor. Chemistry of Materials. 2004;**16**(16):3184-3190. DOI: 10.1021/cm049649j

[28] Zhang LL, Wei T, Wang W, Zhao XS. Manganese oxide–carbon composite as supercapacitor electrode materials. Microporous and Mesoporous Materials. 2009;**123**(1–3):260-267. DOI: 10.1016/j.micromeso.2009.04.008

[29] Xie X, Gao L. Characterization of a manganese dioxide/carbon nanotube composite fabricated using an in situ coating method. Carbon. 2007;**45**(12): 2365-2373. DOI: 10.1016/j.carbon.2007.07.014

[30] Kéranguéven G, Faye J, Royer S, Pronkin SN. Electrochemical properties

and capacitance of Hausmannite Mn_3O_4 – Carbon composite synthesized by in situ autocombustion method. Electrochimica Acta. 2016;**222**:755-764. DOI: 10.1016/j.electacta.2016.11.032

[31] Brousse T, Bélanger D, Long JW, Belanger D, Long JW. To be or not to be pseudocapacitive? Journal of the Electrochemical Society. 2015;**162**(5): A5185-A5189. DOI: 10.1149/2.0201505jes

[32] Levi MD, Aurbach D. Frumkin intercalation isotherm - a tool for the description of lithium insertion into host materials: A review. Electrochimica Acta. 1999;**45**(1):167-185. DOI: 10.1016/S0013-4686(99)00202-9

[33] Levi MD, Aurbach D. Diffusion coefficients of lithium ions during intercalation into graphite derived from the simultaneous measurements and modeling of electrochemical impedance and potentiostatic intermittent titration characteristics of thin graphite electrodes. Journal of Physical Chemistry B. 1997;**101**(23):4641-4647. DOI: 10.1021/jp9701911

[34] Prosini PP, Lisi M, Zane D, Pasquali M. Determination of the chemical diffusion coefficient of lithium in LiFePO4. Solid State Ionics. 2002;**148** (1–2):45-51. DOI: 10.1016/S0167-2738(02)00134-0

[35] Ding N, Xu J, Yao YX, Wegner G, Fang X, Chen CH, et al. Determination of the diffusion coefficient of lithium ions in nano-Si. Solid State Ionics. 2009; **180**(2–3):222-225. DOI: 10.1016/j.ssi.2008.12.015

[36] Delacourt C, Ati M, Tarascon JM. Measurement of lithium diffusion coefficient in Li y $FeSO_4$ F. Journal of the Electrochemical Society. 2011; **158**(6):A741-A749. DOI: 10.1149/1.3581087

[37] Nickol A, Schied T, Heubner C, Schneider M, Michaelis A, Bobeth M, et al. GITT analysis of lithium insertion cathodes for determining the lithium diffusion coefficient at low temperature: Challenges and pitfalls. Journal of the Electrochemical Society. 2020; **167**(9):090546. DOI: 10.1149/1945-7111/ab9404

[38] Park M, Zhang X, Chung M, Less GB, Sastry AM. A review of conduction phenomena in Li-ion batteries. Journal of Power Sources. 2010;**195**(24):7904-7929. DOI: 10.1016/j.jpowsour.2010.06.060

[39] Ghodbane O, Pascal JL, Favier F. Microstructural effects on charge-storage properties in MnO_2-based electrochemical supercapacitors. ACS Applied Materials and Interfaces. 2009; **1**(5):1130-1139. DOI: 10.1021/am900094e

[40] Kerner Z, Pajkossy T. On the origin of capacitance dispersion of rough electrodes. Electrochimica Acta. 2000; **46**(2–3):207-211. DOI: 10.1016/S0013-4686(00)00574-0

[41] Alexander CL, Tribollet B, Vivier V, Orazem ME. Contribution of surface distributions to constant-phase-element (CPE) behavior: 1. Influence of roughness. Electrochimica Acta. 2017; 251:416–424. DOI: 10.1016/j.electacta.2017.08.081; Alexander CL, Tribollet B, Vivier V, Orazem ME. Contribution of surface distributions to constant-phase-element (CPE) behavior: 2. Capacitance. Electrochimica Acta. 2017;251:566–573. DOI: 10.1016/j.electacta.2017.08.081; Alexander CL, Tribollet B, Vivier V, Orazem ME. Contribution of surface distributions to constant-phase-element (CPE) Behavior: 3. Adsorbed intermediates. Electrochimica Acta. 2017;251, 99–108. DOI: 10.1016/j.electacta.2017.08.081

[42] MacDonald DD. Reflections on the history of electrochemical impedance spectroscopy. Electrochimica Acta. 2006;**51**(8–9):1376-1388. DOI: 10.1016/j.electacta.2005.02.107

[43] Keiser H, Beccu KD, Gutjahr MA. Abschätzung der porenstruktur poröser elektroden aus impedanzmessungen. Electrochimica Acta. 1976;**21**(8):539-543. DOI: 10.1016/0013-4686(76)85147-X

[44] Ba H, Wang W, Pronkin S, Romero T, Baaziz W, Nguyen-Dinh L, et al. Biosourced foam-like activated carbon materials as high-performance supercapacitors. Advanced Sustainable Systems. 2018;**1700123**:1700123. DOI: 10.1002/adsu.201700123

[45] Kötz R, Carlen M. Principles and applications of electrochemical capacitors. Electrochimica Acta. 2000; **45**:2483-2498

[46] Eikerling M, Kornyshev AA, Lust E. Optimized structure of nanoporous carbon-based double-layer capacitors. Journal of the Electrochemical Society. 2005;**152**(1):E24-E33. DOI: 10.1149/1.1825379

[47] Pronkin SN, Shokina NY. Generalized staircase model of electrochemical impedance of pores in supercapacitor electrodes. Computational Technologies. 2021; **26**(5):30-51. DOI: 10.25743/ict.2021.26.5.004

Section 3

Molecules and Fundamental Electrochemistry

Chapter 4

Effects of Electrolyte on Redox Potentials

John R. Miller and Matthew J. Bird

Abstract

Redox potentials, especially as measured by cyclic voltammetry and related electrochemical techniques, are the basis for understanding energetics of photochemical solar energy storage, organic photovoltaics, light-emitting diodes, and even photosynthesis. These very popular techniques are dominant although none of the energy systems just mentioned contain large concentrations, typically 100 mM, of supporting electrolyte needed for electrochemical techniques to work. At the same time, the added electrolytes often have large, but unknown effects on the energetics studied. Despite substantial efforts using microelectrodes, it has not been possible to utilize electrochemical techniques to measure redox potentials in the absence of electrolytes. This chapter will be an account of new techniques applying the method of pulse radiolysis to partly answer the question: what is the effect of electrolytes on redox potentials?

Keywords: redox potentials, electrolytes, electron transfer, microelectrodes

1. Introduction

Redox potentials are essential to our understanding of energetics of photosynthesis, organic photovoltaics, dye-sensitized solar cells, redox catalysis, organic light-emitting diodes (LEDs), lithium and flow batteries, and respiratory electron transport cycles. Electrochemical methods such as cyclic voltammetry, polarography, and differential pulse voltammetry are the main sources of redox potentials. But to operate, these techniques require the addition of electrolyte, usually at concentrations near 0.1 M. The presence of electrolyte should stabilize radical cations or radical anion altering redox potentials from their true values in systems mentioned above, many of which contain no electrolyte.

Measurement of the effects of electrolytes on redox potentials has not been easy, but the advent of micro- and ultramicro electrodes enabled several groups to observe electrochemical waves at strongly reduced or even zero electrolyte concentrations [1–13]. Still these waves are difficult to interpret to obtain values of $\Delta E^{o'}$, the difference between the redox potential with, $E^{o'}_{[XY]\,=\,0.1}$ and without, $E^{o'}_{[XY]\,=\,0}$ electrolyte,

$$\Delta E^{o'} = E^{o'}_{[XY]=0} - E^{o'}_{[XY]=0.1} \qquad (1)$$

Here XY is an inert, supporting electrolyte. A popular electrolyte is tetrabutylammonium hexafluorophosphate, $TBAPF_6$. In our reading of the literature, we have yet to find even one report of the difference $\Delta E^{o'}$ for 0.1 moles/l (M) or any other concentration of electrolyte by electrochemical methods.

2. What is the effect of electrolytes on redox potentials?

While reports of the change in redox potential due to addition of electrolyte are rare, some results from our laboratory became available recently [14, 15]. To introduce the concept of shifts of redox potentials, **Figure 1** displays potentials for four couples, t-stilbene$^{0/-}$, F_{10} $^{0/-}$, $Cocp_2$$^{+/0}$, and $Fecp_2$$^{+/0}$ (referred to herein as Fc) in THF over a wide range of concentrations of the electrolyte $TBAPF_6$. F_{10} is a ten-unit oligomer of 9,9-dihexyl fluorene. Results for Stilbene and F_{10} used only experimental data. Those for $Cocp_2$ and $Fecp_2$$^{+/0}$ used experiment and computations and may thus be slightly less reliable.

Figure 1 displays the potentials of the four species vs. the reference of an electron in vacuum (left axis) or the aqueous saturated calomel electrode (right axis) SCE. These two external references remain constant as electrolyte is added. **Table 1**, below, on the other hand, uses the internal ferrocene ($Fecp_2$) as reference, while keeping the reference constant by always using it in 0.1 M $TBAPF_6$. Methods for establishing and comparing reference potentials are described in the section Reference couples and electrodes.

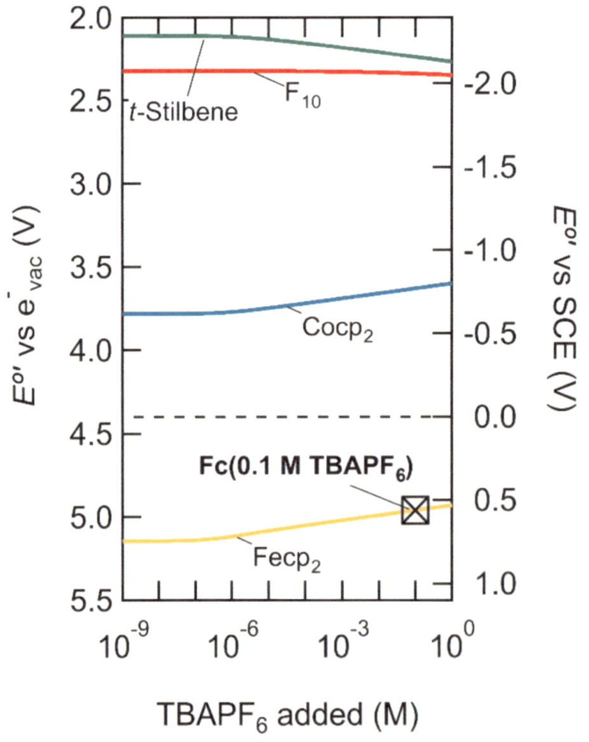

Figure 1.
The variation of redox potentials with the concentration of electrolyte, $TBAPF_6$, from $[TBAPF_6]$ = 0 to 1.0 M. The position of $E^{o'}$ (Fc) is marked.

	Electrolyte[a]	K_d (M)[b]	$E^{o'}$ vs. $E^{o'}_{[[XY]=0.1]}$[c]	$E^{o'}_{[[XY]=0]}$[c]	ΔE mV	Ref
$Stilbene^{0/-}$	$TBAPF_6$	$3.1x10^{-6}$	−2.72	−2.85	129	[14]
$F_{10}{}^{0/-}$	$TBAPF_6$	$1.8x10^{-3}$	−2.62	−2.64	17	[14]
1-MePyrene	$TBAPF_6$	$4.2x10^{-6}$	−2.60	−2.73	128	[14]
$Cocp_2{}^{+/0}$	$TBAPF_6$	$1.4x10^{-6}$	−1.33	−1.48	151	b
$Fecp_2{}^{+/0}$	$TBAPF_6$	$4.1x10^{-7}$	0	0.183	183	b
$Per^{0/-}$	$TBAPF_6$	5.9×10^{-7}	−2.21	−2.38	174	[15]
$BzPh^{0/-}$	$TBAPF_6$	6.3×10^{-8}	−2.32	−2.55	232	[15]
$Per^{0/-}$	$NaBPh_4$	2.3×10^{-5}	−2.26	−2.38	124	[15]
$BzPh^{0/-}$	$NaBPh_4$	6.8×10^{-11}	−2.10	−2.55	451	[15]
$Per^{0/-}$	$\{Na\}BPh_4$	6.4×10^{-5}	−2.28	−2.38	100	[16]
$BzPh^{0/-}$	$\{Na\}BPh_4$	2.2×10^{-5}	−2.38	−2.55	173	[16]

[a]*Electrolytes are tetrabutylammonium hexafluorophosphate ($TBAPF_6$, $K_d = 2.7 \times 10^{-6}$ M, [17]) sodium tetraphenylborate ($NaBPh_4$, $K_d = 8.8x10^{-5}$ M [18]) and sodium tetraphenylborate with the Na^+ encapsulated in the 222-cryptand ($\{Na\}BPh_4$, $K_d = 9.3.x10^{-5}$ M [15]). MePy is 1-methylpyrene, Per is perylene, BzPh is benzophenone.*
[b]*K_d for ($Cocp_2{}^+$, $PF_6{}^-$) and ($Fecp_2{}^+$, $PF_6{}^-$) were estimated from measured [14] K_d's for Stilbene and MePyrene and the ratios of computed K_d's for those to computed K_d's for ($Cocp_2{}^+$, $PF_6{}^-$) and ($Fecp_2{}^+$, $PF_6{}^-$).*
[c]*$E^{o'}$ is the reduction potential in 0.1 M electrolyte vs. Fc in THF with 0.1 M $TBAPF_6$. E^o is the reduction potential without electrolyte, also vs. Fc in THF with 0.1 M $TBAPF_6$. Potentials vs. Fc in THF with no electrolyte are more negative by 0.185 V.*

Table 1.
Dissociation constants, K_d, and redox potentials in THF. Redox potentials are vs. Fc ($Fecp_2{}^{+/0}$) with 100 mM $TBAPF_6$.

3. Redox potential shifts by ion pairing and activities

When electrolyte is added, two factors shift redox potentials: 1) ion pairing and 2) activity.

Both are pictured in **Figure 2** and included in Eq. (2), below. In **Figure 2**, one positively charged ion of the electrolyte is close to the radical anion, forming an ion pair. Other ions of the electrolyte contribute to the ionic atmosphere. In highly polar media such as water or acetonitrile, there is little ion pairing and most electrolyte exists as free ions. Ionic atmospheres around radical anions or cations stabilize them, described as change of their activities, γ, calculated [14] here using extended Debye-Hückel theory. In these calculations in redox potential, the component attributed to activities is usually small. For 100 mM $TBAPF_6$ estimates of the shift from activities

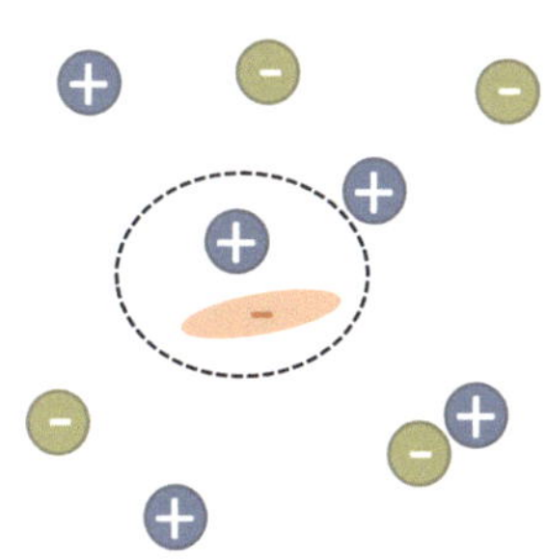

Figure 2.
A radical ion (red oval) in solution surrounded by + and – ions of an electrolyte.

are 18 mV in THF and 24 mV in acetonitrile using K_d's for $TBAPF_6$ from LeSuer, Buttolph, and Geiger [17].

Shifts of redox potentials in **Figure 1**, calculated by Eq. (2), are much larger. The data in **Figure 1** are in THF, where ion pairing can produce large changes in redox potentials.

$$E^{o'}_{[XY]} - E^{o}_{[XY]=0} = \frac{RT}{F} \ln\left(\frac{1}{\gamma_\pm} + \frac{\gamma_\pm [X^+]}{K_d^o(M^{\bullet -}, X^+)}\right) \quad (2)$$

$$E^{o'}_{[XY]} - E^{o}_{[XY]=0} = -\frac{RT}{F} \ln\left(\frac{1}{\gamma_\pm} + \frac{\gamma_\pm [Y^-]}{K_d^o(M^{\bullet +}, Y^-)}\right) \quad (3)$$

Eq. (2), derived in Ref., [14] gives the change in redox potential with concentration of the electrolyte, XY, in terms of a dissociation constant, K_d, and an activity coefficient, $\gamma_\pm$. The free electrolyte concentrations, $[X^+]$ or $[Y^-]$, are found using the dissociation constants for the electrolyte in THF. It is derived by applying the Nernst equation in the presence of an ion-pairing equilibrium. Reference [14] also presents data used to obtain the parameters needed in Eq. (2). That data also examines and supports the model of Eq. (2). Eq. (2) is for reduction of molecule M to $M^{\cdot -}$ (e.g., Stilbene$^{\cdot -}$; (3) is for reduction of $M^{\cdot +}$ (e.g. $Cocp_2^{\cdot +}$) to M. Qu and Persson [19] gave an equation similar to Eq. (2), although without effects of activity.

Ion pairing is strong in media of lower polarity. The shifts of redox potentials in **Figure 1** and **Table 1** (below) are in THF, a liquid of moderately low polarity with a dielectric constant ε = 7.6. With this low polarity, Eq. (2) predicts the ion-pairing contribution to shift redox potentials by amounts that are generally much larger than that due to activity (~18 mV). A notable exception is the case of the highly delocalized anion of F_{10}, to be discussed below. This is especially evident for the case of the largest shift, 451 mV for benzophenone radical anion paired with Na^+. In that case, the small Na^+ ion pairs strongly with the negative charge concentrated on the ketone of benzophenone. The shift is smaller by almost a factor of 2 for pairing with TBA^+ and still smaller with Na^+ encapsulated in the 2.2.2. cryptand.

4. Methods to measure redox potentials with and without electrolyte

While we yet look forward to the possibility of measurements of redox potentials without electrolyte by electrochemistry at micro and ultramicro electrodes, two methods using pulse radiolysis have emerged from our laboratory. Both measure electron transfer equilibria without electrolyte and with electrolyte, studying electron transfer from a radical anion, $D^{\bullet -}$ to an acceptor, A, under conditions where the equilibrium can be observed and the equilibrium constant, K_{eq}, measured. Creating $D^{\bullet -}$ by pulse radiolysis (**Figure 3**) enables determination of K_{eq}.

In **Figure 3**, ionizations create electrons and holes, which add to solutes to create radical ions. In some liquids, such as THF, only radical anions are created. Radical cations can be created in other liquids. When an inert electrolyte is added, the electron transfer equilibrium is coupled to two ion-pairing equilibria as shown in **Figure 4**.

4.1 Method 1

If the two ion-pairing equilibria have the same dissociation constants, $K_{dd} = K_{da}$, then the apparent "composite" equilibrium constant, K_{eqC}, does not shift as electrolytes are added. If $D^{\bullet -}$ and $A^{\bullet -}$ have different K_d's for pairing with counter ions of the

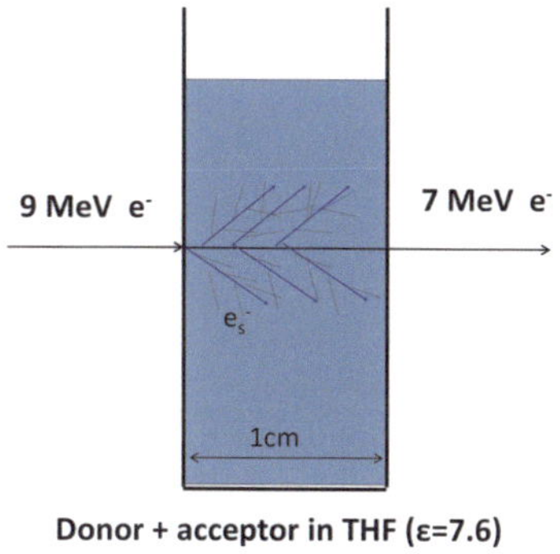

Figure 3.
A spectrophotometric cell containing a solution. In our accelerator, LEAF, 9 MeV electrons enter one side of the cell creating ionizations in the solution. Each ionization is followed by secondary and tertiary ionizations.

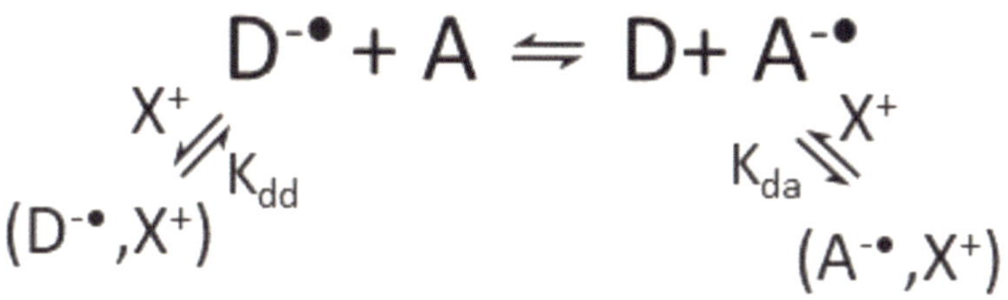

Figure 4.
An electron transfer equilibrium coupled to two ion-pairing equilibria with inert ion X^+. K_{dd} and K_{da} are dissociation constants for the ion pairs of $D^{\bullet-}$ and $A^{\bullet-}$ with the counter-ion of the electrolyte.

electrolyte, then K_{eqC}, shifts as electrolyte is added. K_{eqC} is the equilibrium constant between all D- species and all A- species. The system of three coupled equilibria has a simple, analytic solution under appropriate conditions, enabling the measurements of as a function of electrolyte concentration to be fit to yield K_{eq}, K_{dd}, and K_{da}. Eq. (2) then gives the shifts of redox potentials for both $D^{0/-}$ and $A^{0/-}$ [14]. The good fit to the equations describing this mechanism also serve to validate the mechanism.

This method is very general requiring only that K_{dd} and K_{da} are substantially different. A difference of a factor of 10 is sufficient. It also needs either $D^{\bullet-}$ or $A^{\bullet-}$ to have a readily measurable absorption spectrum. An aspect of the generality of this method is that it enables determination of dissociation constants of radical ions with *any* counter ion, e.g., TBA^+. **Figure 5** presents an example of data using this method.

Previously, Szwarc determined K_d's for pairs like (biphenyl$^{\bullet-}$, Na^+) by conductivity, but the prior methods were applicable only to alkali metal ions. Reports in **Table 1** of example like K_d(stilbene$^{\bullet-}$, TBA^+) in Ref. [14] are new. The method can be readily generalized to equilibria of radical cations.

4.2 Method 2

A second method uses thermodynamic cycles like that in **Figure 6**. This cycle was constructed from measured equilibrium constants in THF for electron transfer from perylene to benzophenone,

$$\mathrm{Per}^{\bullet-} + \mathrm{BzPh} \rightleftharpoons \mathrm{Per} + \mathrm{BzPh}^{\bullet-} \qquad (4)$$

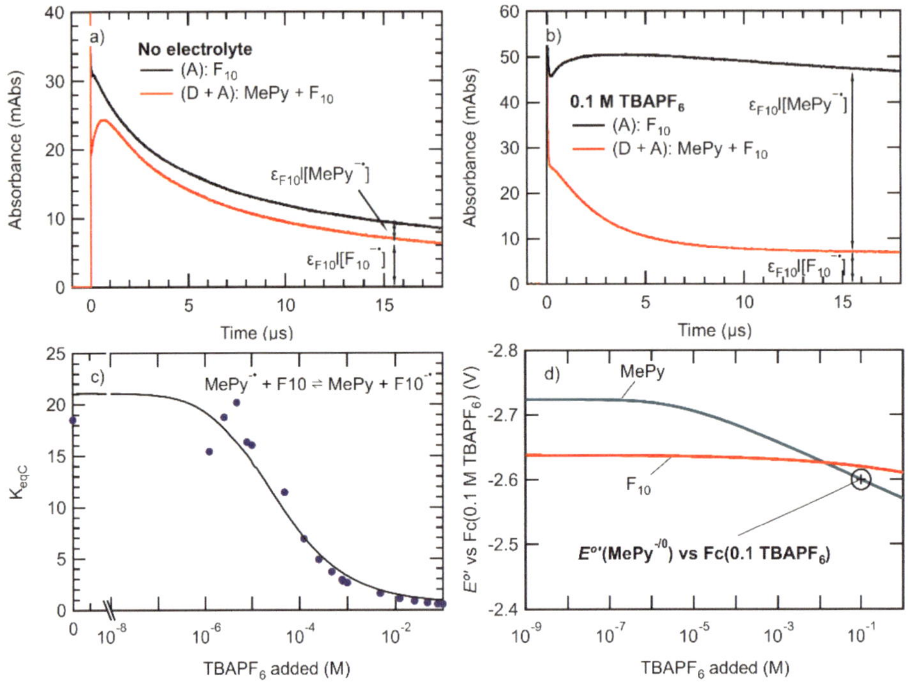

Figure 5.
Measurement of KeqC for electron transfer from $MePy^{\bullet-}$ *to* F_{10} *as a function of electrolyte concentration. Transient absorption at 1574 nm where only* $F_{10}^{\bullet-}$ *absorbs in THF a) without electrolyte and b) with 0.1 M* $TBAPF_6$*. c)* K_{eqC} *from the transient absorption data and d) the redox potentials for MePy and* F_{10}*.*

In **Figure 6**, the top segment is the measured free energy change for reaction [4] without electrolyte; the bottom is the free energy change with $NaBPh_4$ added. Strong ion pairing in ($BzPh^{\cdot-}$, Na^+) changes the reaction from endoergic (ΔG = +171 meV) to exoergic (ΔG = -155 meV) in the presence of sodium. K_d($Per^{\cdot-}$, Na^+) measured by Slates, and Szwarc [20] specifies the left segment, while the cycle yields the right segment, giving K_d($BzPh^{\cdot-}$, Na^+) = 6.8 x 10^{-11} M.

The measurement of this the very small K_d($BzPh^{\cdot-}$, Na^+) illustrates the power of Method 2 for select cases, although this method can work only when external data supplies one segment of the cycle, in this case K_d($Per^{\cdot-}$, Na^+) measured by conductivity [20], which gives the left leg of the cycle.

5. Reference couples and electrodes

IUPAC recommended reporting of redox potentials relative to the internal ferrocenium/ferrocene reference (Fc) [21]. Our measurements to obtain redox potentials in THF at small or zero electrolyte concentrations [14, 15] used the determinations of Shalev and Evans [22] to reference our measurements to Fc in THF with 100 mM $TBAPF_6$. Potentials in **Table 1** use this reference, which we will call Fc(THF, $TBAPF_6$). Common reference electrodes are SCE, the standard hydrogen electrode, silver ion (Ag), Ag/AgCl, and internal references. In **Figure 1**, we referred potentials to the electron in vacuum and aqueous SCE because these do not shift with electrolyte concentration. Here we applied Connelly and Geiger's finding that Fc(THF, $TBAPF_6$) is 0.56 V vs. SCE [23] and references for vs. e- in vacuum [24, 25].

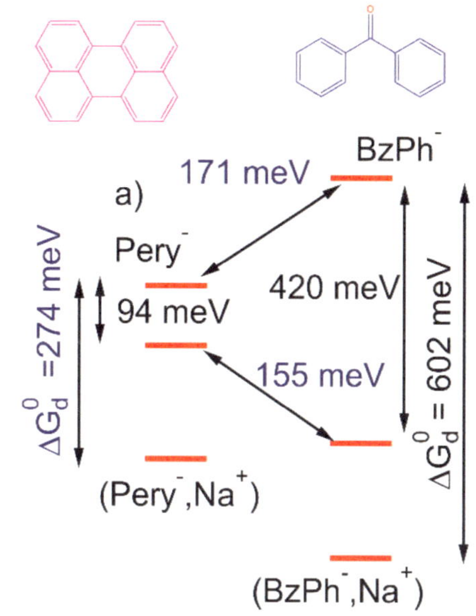

Figure 6.
Free energy cycle for electron transfer from Per$^{\bullet -}$ *to BzPh in THF without and with* $NaBPh_4$ *electrolyte.*

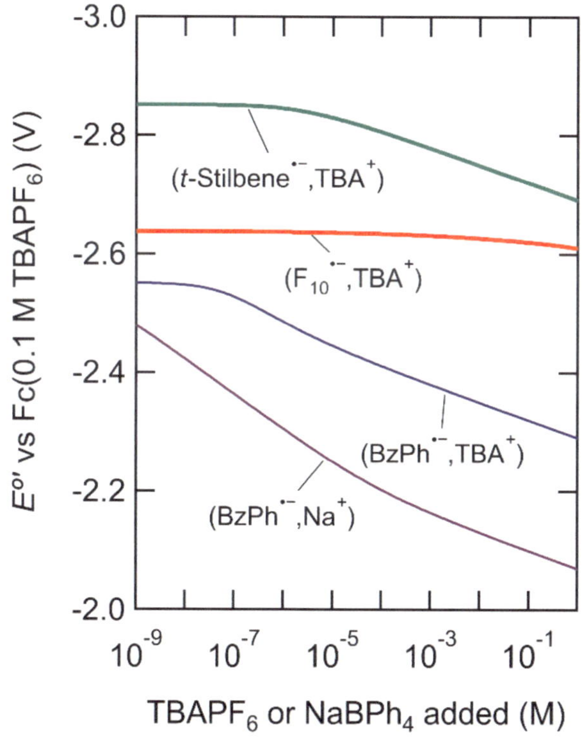

Figure 7.
Shifts of redox potentials for reduction in THF for benzophenone (BzPh), t-stilbene and F_{10} *vs. the fixed reference of Fc(100 mM* $TBAPF_6$*).*

6. Redox and reference shifts

Figures 5 and 7 graph redox potentials with two difference references.

In **Figure** 7, we see that the redox potential of $BzPh^{0/-}$ increases more rapidly with $TBAPF_6$ concentration than either that Stilbene or F_{10}. In the presence of Na^+ ions, the redox potential increases still more rapidly and the increase begins even before the 10^{-9} M limit in the graph.

In **Figure** 7, the redox potential of $F_{10}{}^{0/-}$ changes little due to the remarkably large $K_d(F_{10}{}^{\bullet-}, TBA^+) = 1.8x10^{-3}$ M. The weak association in this pair is due to weak electrostatic attraction of the positively charged TBA^+ ion to the highly delocalized charge distribution of $F_{10}{}^{\bullet-}$ in the ($F_{10}{}^{\bullet-}$, TBA^+) pair. **Figure 8** shows the computed ion pair using density functional theory (b3lyp/6–31 + g*) in Gaussian 16 [27] with the SMD solvation model for [28] THF.

The redox potential of $F_{10}{}^{0/-}$ changes more in **Figure 9** where the reference is Fc at each electrolyte concentration. Most of this change is *not* due to changes in the potential of $F_{10}{}^{0/-}$, but to changes in the Fc reference potential with increasing $TBAPF_6$ concentration. Vullev and coworkers emphasized the effect of electrolyte on the potential of ferrocene [29].

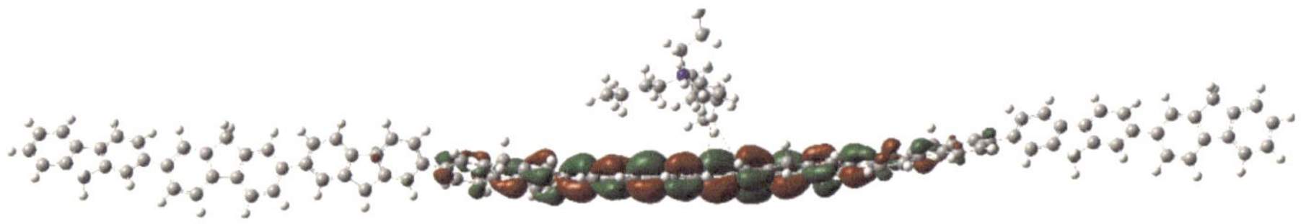

Figure 8.
Pairing of the TBA^+ ion with the very delocalized charge in the anion of F_{10} rendered by Gaussview [26].

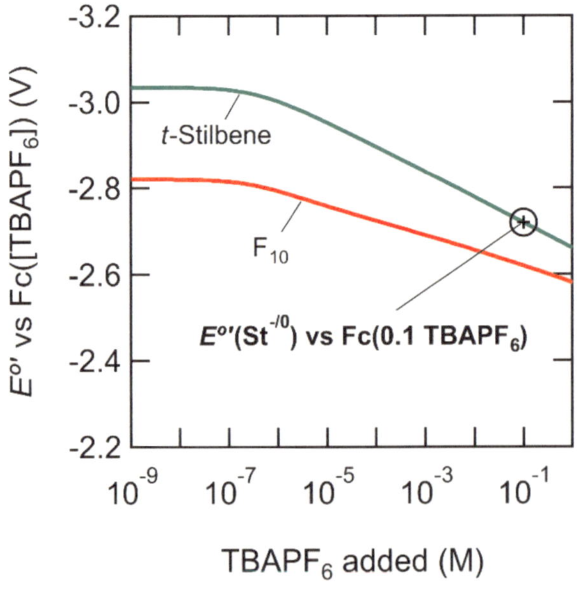

Figure 9.
Shifts of redox potentials for reduction in THF of t-stilbene and F_{10} vs. the variable reference of fc[$TBAPF_6$]; at each concentration [$TBAPF_6$] the reference is Fc at that concentration. The position of $E^{o'}$ (Stilbene) is marked.

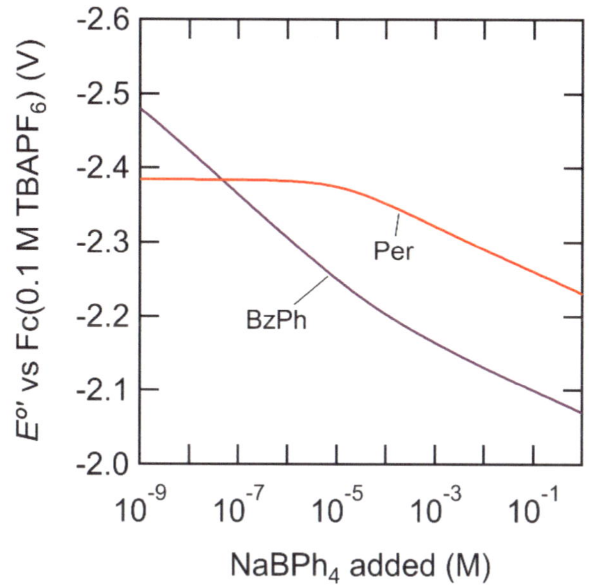

Figure 10.
Redox potentials of BzPh and per vs. $NaBPh_4$ concentration.

In **Figure 10**, the redox potentials of benzophenone (BzPh) and Perylene (Per) depend differently on the $NaBPh_4$ electrolyte concentration due to very different K_d's (**Table 1**), but the two curves are almost parallel at large [$NaBPh_4$]; the difference between the two changes by only 11 mV from [$NaBPh_4$] = 10^{-4} M to 0.1 M. This constancy occurs because changes in redox potential due to ion pairing are constant at 59 mV per decade of counter ion concentration once the onset of the effect occurs.

7. Conclusions and prospects

The effect of electrolytes on redox potentials is now becoming known. Those effects are expected to be small in polar media where ion pairing is not strong and very large in nonpolar media where ion pairing is expected to be very strong. This chapter describes effects of electrolytes in a liquid of medium polarity, THF (ε = 7.6), in which the measurements are moderately challenging, but possible using new methods based on pulse radiolysis.

A principal contributor to redox potentials dissociation constants, K_d, for radical ions paired with counterions from the electrolyte. Few of these are known in the literature, but this work also reported several of these. The two methods described above used pulse radiolysis to determine K_d's from 6.8 x 10^{-11} to 1.8 x 10^{-3} M.

In addition, these pulse radiolysis methods will enable testing of new internal references and "weakly coordinating" anions [30, 31] and cations [32], which could lead to better candidates for use in electrochemistry, particularly in low-polarity solvents. Furthermore, these methods enable the measurement of referenced redox potentials in the absence of electrolyte and can go beyond typical electrochemical potential windows [33].

Acknowledgements

This material is based upon work supported by the U.S. Department of Energy, Office of Science, Office of Basic Energy Sciences, Division of Chemical Sciences, Geosciences & Bioscience, through Grant DE-SC0012704, including use of the LEAF and Van de Graaff facilities of the BNL Accelerator Center for Energy Research. We are grateful to Tomokazu Iyoda, Matthew Pearson, Abram Ledbetter, and Nick Bonura for their contributions to references [14, 15].

Appendices and nomenclature

$E^{o'}_{[XY]\,=\,0.1}$	is the redox potential in 0.1 M of electrolyte, XY.
$E^{o'}_{[XY]\,=\,0.0}$	is the redox potential without electrolyte.
$\Delta E^{o'}$	is the difference between the redox potential with and without electrolyte,
K_{eqC}	is the composite equilibrium constant for an electron transfer coupled to two ion pairing equilibria (see **Figure 4**).
M	is the concentration in moles/l.

Author details

John R. Miller* and Matthew J. Bird
Chemistry Division, Brookhaven National Laboratory, Upton, NY, USA

*Address all correspondence to: jrmiller@bnl.gov

References

[1] Bond AM, Fleischmann M, Robinson J. Electrochemistry in organic-solvents without supporting electrolyte using platinum microelectrodes. Journal of Electroanalytical Chemistry. 1984;**168** (1–2):299-312. DOI: 10.1016/0368-1874(84)87106-3

[2] Bond AM, Lay PA. Cyclic voltammetry at microelectrodes in the absence of added electrolyte using a platinum quasi-reference electrode. Journal of Electroanalytical Chemistry. 1986;**199**(2):285-295. DOI: 10.1016/0022-0728(86)80004-3

[3] Cooper JB, Bond AM. Microelectrode studies in the absence of deliberately added supporting electrolyte - solvent dependence for a neutral and singly charged species. Journal of Electroanalytical Chemistry. 1991;**315**(1–2):143-160. DOI: 10.1016/0022-0728(91)80066-y

[4] Bond AM, Pfund VB. Cyclic voltammetry at gold, platinum and carbon microelectrodes in ice without added supporting electrolyte - evidence for liquid microphases at temperatures well below the freezing-point of water. Journal of Electroanalytical Chemistry. 1992;**335**(1–2):281-295. DOI: 10.1016/0022-0728(92)80248-3

[5] Cooper JB, Bond AM, Oldham KB. Microelectrode studies without supporting electrolyte - model and experimental comparison for singly and multiply charged ions. Journal of Electroanalytical Chemistry. 1992;**331**(1–2):877-895. DOI: 10.1016/0022-0728(92)85012-r

[6] Cooper JB, Bond AM. Evidence for adsorption of the Cobaltocenium cation and precipitation of uncharged Cobaltocene at the platinum microelectrode acetonitrile Interface in the absence of supporting electrolyte. Analytical Chemistry. 1993;**65**(20): 2724-2730. DOI: 10.1021/ac00068a004

[7] Heinze J. Ultramicroelectrodes in electrochemistry. Angewandte Chemie-International Edition in English. 1993; **32**(9):1268-1288. DOI: 10.1002/anie.199312681

[8] Amatore C, Paulson SC, White HS. Successive electron-transfers in low ionic strength solutions. Migrational flux coupling by homogeneous electron transfer reactions. Journal of Electroanalytical Chemistry. 1997; **439**(1):173-182. DOI: 10.1016/s0022-0728(97)00382-3

[9] Oldham KB, Cardwell TJ, Santos JH, Bond AM. Effect of ion pairing on steady-state voltammetric limiting currents at microelectrodes. 1. Theoretical principles. Journal of Electroanalytical Chemistry. 1997;**430**(1–2):25-37. DOI: 10.1016/s0022-0728(96)04914-5

[10] Oldham KB, Cardwell TJ, Santos JH, Bond AM. Effect of ion pairing on steady state voltammetric limiting currents at microelectrodes. 2. Experimental studies on charged (Br-, Ag+) and uncharged (copper diethyldithiocarbamate) species in toluene. Journal of Electroanalytical Chemistry. 1997;**430**(1–2):39-46. DOI: 10.1016/s0022-0728(96)04915-7

[11] Bento MF, Thouin L, Amatore C, Montenegro MI. About potential measurements in steady state voltammetry at low electrolyte/analyte concentration ratios. Journal of Electroanalytical Chemistry. 1998; **443**(1):137-148. DOI: 10.1016/s0022-0728(97)00459-2

[12] Silva SM, Bond AM. Contribution of migration current to the voltammetric

deposition and stripping of lead with and without added supporting electrolyte at a mercury-free carbon fibre microdisc electrode. Analytica Chimica Acta. 2003; **500**(1–2):307-321. DOI: 10.1016/s0003-2670(03)00881-x

[13] Limon-Petersen JG, Dickinson EJF, Belding SR, Rees NV, Compton RG. Cyclic voltammetry in weakly supported media the reduction of the cobaltocenium cation in acetonitrile - comparison between theory and experiment. Journal of Electroanalytical Chemistry. 2010;**650**(1):135-142. DOI: 10.1016/j.jelechem.2010.08.011

[14] Bird MJ, Pearson MA, Asaoka S, Miller JR. General method for determining redox potentials without electrolyte. Journal of Physical Chemistry A. 2020;**124**(26):5487-5495. DOI: 10.1021/acs.jpca.0c02948

[15] Bird MJ, Iyoda T, Bonura N, Bakalis J, Ledbetter AJ, Miller JR. Effects of electrolytes on redox potentials through ion pairing. Journal of Electroanalytical Chemistry. 2017;**804**: 107-115. DOI: 10.1016/j.jelechem.2017.09.030

[16] An alternative path from Fc(THF, TBAPF6) to e- vac uses Shalev's report that aqueous SCE is 0.93 V vs. the potential of Cocp2 in acetonitrile (MeCN) with 100 mM TBAPF6. A conversion using computed solvation energies places SCE at 0.704 vs. Cocp2 (THF,TBAPF6). They find Fc(THF, TBAPF6) to be 1.332 V vs. Cocp2(THF, TBAPF6) placing Fc(THF,TBAPF6) 5.12 V vs e-. We used the average of these two

[17] LeSuer RJ, Buttolph C, Geiger WE. Comparison of the conductivity properties of the Tetrabutylammonium salt of Tetrakis(pentafluorophenyl) borate anion with those of traditional supporting electrolyte anions in nonaqueous solvents. Analytical Chemistry. 2004;**76**(21):6395-6401. DOI: 10.1021/ac040087x

[18] Nicholls D, Sutphen C, Szwarc M. Dissociation of lithium and sodium salts in ethereal solvents. The Journal of Physical Chemistry. 1968;**72**(3): 1021-1027. DOI: 10.1021/j100849a041

[19] Qu X, Persson KA. Toward accurate modeling of the effect of ion-pair formation on solute redox potential. Journal of Chemical Theory and Computation. 2016;**12**(9):4501-4508. DOI: 10.1021/acs.jctc.6b00289

[20] Slates RV, Szwarc M. Dissociative equilibria in systems aromatic hydrocarbon- Na+— – radical anion- + Na+. Journal of Physical Chemistry. 1965;**69**(12):4124. DOI: 10.1021/j100782a012

[21] Gritzner G, Kuta J. Recommendations on reporting electrode-potentials in nonaqueous solvents. Pure and Applied Chemistry. 1984;**56**(4):461-466. DOI: 10.1351/pac198456040461

[22] Shalev H, Evans DH. Solvation of anion radicals: Gas-phase versus solution. Journal of the American Chemical Society. 1989;**111**(7): 2667-2674. DOI: 10.1021/ja00189a048

[23] Connelly NG, Geiger WE. Chemical redox agents for organometallic chemistry. Chemical Reviews. 1996; **96**(2):877-910. DOI: 10.1021/cr940053x

[24] Cardona CM, Li W, Kaifer AE, Stockdale D, Bazan GC. Electrochemical considerations for determining absolute frontier orbital energy levels of conjugated polymers for solar cell applications. Advanced Materials. 2011;

23(20):2367-2371. DOI: 10.1002/adma.201004554

[25] Isse AA, Gennaro A. Absolute potential of the standard hydrogen electrode and the problem of interconversion of potentials in different solvents. The Journal of Physical Chemistry B. 2010;**114**(23):7894-7899. DOI: 10.1021/jp100402x

[26] Dennington R, Keith TA, Millam JM. GaussView, Version 6.0. Shawnee Mission, KS: Semichem Inc.; 2016

[27] Frisch MJ, Trucks GW, Schlegel HB, Scuseria GE, Robb MA, Cheeseman JR, et al. Gaussian 16, Revision B.01. Wallingford CT: Gaussian, Inc.; 2016

[28] Marenich AV, Cramer CJ, Truhlar DG. Universal solvation model based on solute electron density and on a continuum model of the solvent defined by the bulk dielectric constant and atomic surface tensions. The Journal of Physical Chemistry B. 2009;**113**(18): 6378-6396. DOI: 10.1021/jp810292n

[29] Bao D, Millare B, Xia W, Steyer BG, Gerasimenko AA, Ferreira A, et al. Electrochemical oxidation of ferrocene: A strong dependence on the concentration of the supporting electrolyte for nonpolar solvents. Journal of Physical Chemistry A. 2009;**113**(7): 1259-1267. DOI: 10.1021/jp809105f

[30] Khan FST, Waldbusser AL, Carrasco MC, Pourhadi H, Hematian S. Synthetic, spectroscopic, structural, and electrochemical investigations of ferricenium derivatives with weakly coordinating anions: Ion pairing, substituent, and solvent effects. Dalton Transactions. 2021;**50**(21):7433-7455. DOI: 10.1039/D1DT01192H

[31] Geiger WE, Barriere F. Organometallic electrochemistry based on electrolytes containing weakly-coordinating Fluoroarylborate anions. Accounts of Chemical Research. 2010; **43**(7):1030-1039. DOI: 10.1021/ar1000023

[32] Mann L, Hornberger E, Steinhauer S, Riedel S. Further development of weakly coordinating cations: Fluorinated Bis (triarylphosphoranylidene)iminium salts. Chemistry – A European Journal. 2018;**24**(15):3902-3908. DOI: 10.1002/chem.201705992

[33] Bird MJ, Cook AR, Zamadar M, Asaoka S, Miller JR. Pushing the limits of the electrochemical window with pulse radiolysis in chloroform. Physical Chemistry Chemical Physics. 2020; **22**(26):14660-14670. DOI: 10.1039/d0cp01948h

Chapter 5

Charge Carriers for Next-Generation Redox Flow Batteries

Catherine L. Peake, Graham N. Newton and Darren A. Walsh

Abstract

Increasing the volumetric energy density of redox flow batteries beyond that of the archetypal all-vanadium system requires the development of highly soluble charge carriers that can store multiple electrons per charge cycle. In this review article we will describe the design and performance of a range of new charge carriers for flow batteries, with an emphasis on those with multi-electron redox properties. These include fullerene derivatives, multifunctional organic systems, metal coordination complexes, and polyoxometalates. Our discussion will include an evaluation of the fundamental physical and electrochemical properties of the charge carriers and their impact on battery performance and energy density.

Keywords: multi-electron charge carriers, flow battery, electrolyte, next-generation batteries, energy density

1. Introduction

High-capacity, cost effective and durable electrochemical energy storage technologies are necessary to satisfy the growing uptake of intermittent renewable sources and maintain stable electrical-grid systems [1]. In this regard, redox flow batteries (RFBs) are ideally suited and have attracted considerable attention. RFBs differ from conventional batteries in that redox-active molecules, termed charge carriers, are dissolved into electrolyte and are stored in reservoirs external to the electrochemical cell. The charge carrier-containing electrolyte is pumped through electrodes in the electrochemical cell to charge/discharge the battery. Upon charge, energy is stored as the positive electrolyte (also termed catholyte or posolyte) is oxidised and the negative electrolyte (also termed anolyte or negolyte) is simultaneously reduced. Upon discharge, energy is released as the redox reactions are reversed. The electrolytes are separated in the cell by a membrane or separator, which allows transfer of charge-balancing counterions but prevents crossover of charge carriers to the opposite half-cell, thus preventing self-discharge. **Figure 1** shows a schematic of a generic RFB under discharge conditions.

The distinctive design of RFBs allows decoupling of energy and power, and therefore facile scale-up to high capacities [3]. Capacity can be enhanced by simply

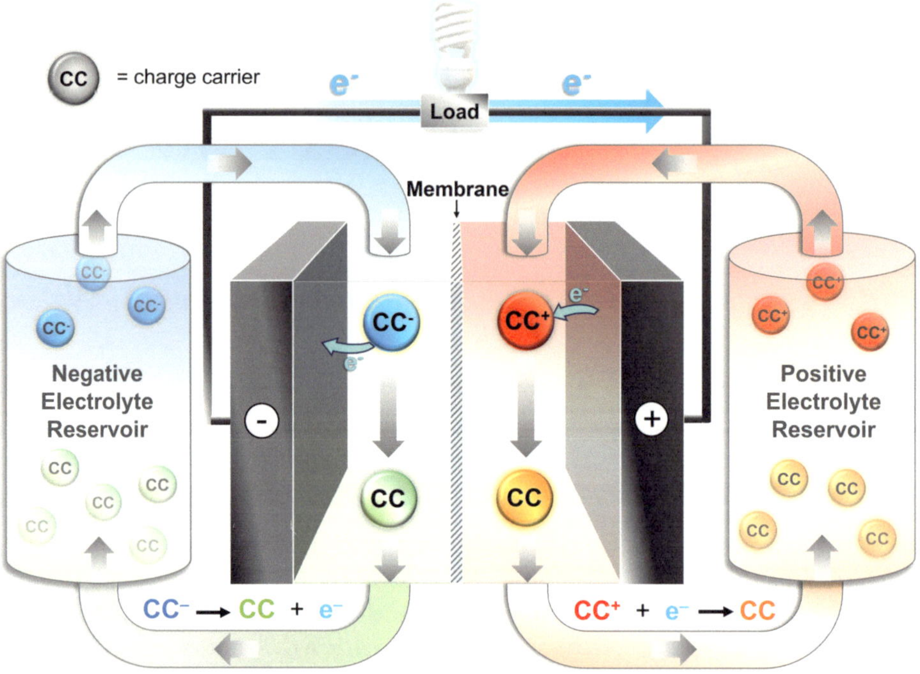

Figure 1.
Schematic of a generic RFB under discharge conditions. Electrolytes are stored in reservoirs and are pumped through the electrochemical cell to charge and discharge the battery. Upon discharge the charge carriers in the positive electrolyte are reduced while those in the negative electrolyte are oxidised. Polarity is reversed to allow the opposite redox reactions to occur upon charge. Reproduced from [2] with permission from the Royal Society of Chemistry.

increasing the volume of electrolyte in the reservoirs, without the need to modify the electrochemical cell, while power is determined by the cell design (e.g. electrode surface area etc). Furthermore, in contrast to conventional batteries such as the lithium-ion battery (LIB), RFBs avoid the intercalation and deintercalation of redox-active molecules between the electrolyte and solid electrode material. Instead, redox reactions occur via solution phase charge carriers at the electrode surface. This underpins the long operational lifetimes of RFBs (15–20 years), making them particularly suited to grid-scale energy storage.

Historically, RFBs have relied on charge carriers based on metals such as iron, chromium, zinc, or cerium, dissolved in aqueous electrolyte [4]. The earliest investigations were conducted by the National Aeronautics and Space Administration (NASA) in the 1970s. Their most notable development was the iron-chromium RFB which used Fe^{2+}/Fe^{3+} and Cr^{2+}/Cr^{3+} in the positive and negative electrolytes respectively [5]. Scale-up and commercialisation of the system were hindered by several technical challenges including the slow electron transfer kinetics of Cr^{2+}/Cr^{3+} [4]. Today, the most commercially advanced RFB system is the symmetric, all-vanadium RFB developed by Skyllas-Kazacos and co-workers in the late 1980s [6, 7]. The charge carriers in the system are VO_2^+/VO^{2+} (V^{5+}/V^{4+}) in the positive electrolyte and V^{2+}/V^{3+} in the negative electrolyte. Vanitec lists 33 companies commercialising all-vanadium RFBs [8] and several plants have been installed globally. The largest electrochemical energy storage plant in the world is forecast to be a 200 MW/800 MWh all-vanadium RFB and is under construction by Ronge Power of China [3].

Despite their advantages for grid-scale energy storage, commercial uptake of the all-vanadium RFB is dwarfed by that of LIBs due to several drawbacks. Firstly, the cost of the all-vanadium RFB was estimated at \$500 kWh^{-1} in 2014 [9], which far exceeds the target of \$100 kWh^{-1} set by the US Department of Energy [10], and the ever decreasing cost of LIBs estimated at \$156 kWh^{-1} in 2020 [11]. Secondly, the energy density of the all-vanadium RFB is an order of magnitude lower than LIBs [2]. The limited solubility of vanadium sulphate in aqueous solution and the cell voltage of approximately 1.3 V (dictated by the difference in redox potential between the reaction at the positive and negative electrode), limits the energy density to 25–35 Wh L^{-1} [12]. While lower energy densities are generally more tolerable for stationary rather than portable applications, there is a demand to enhance RFB energy density to cut cost, reduce space requirements and access new markets.

Energy density is a measure of the energy output per unit volume of total electrolyte and is defined in Eq. (1);

$$\text{Energy density} = \frac{n\,V_{\text{cell}}\,C\,F}{2} \tag{1}$$

where n is the number of electrons transferred per molecule in the charge/discharge redox reaction, V_{cell} is the average cell voltage, C is the concentration of charge carrier in the electrolyte, and F is the Faraday constant. The division by two accounts for the necessity for two volumes of electrolyte for a given energy output (positive and negative electrolyte). Furthermore, fast electron transfer kinetics and high stability of the charge carrier are crucial to achieve high power density and long cycle and calendar life. In this regard, charge carriers underpin the energy density and performance of RFBs, and strongly influences the overall cost and sustainability too.

As indicated in Eq. (1), an effective strategy to enhance the energy density of RFBs is to increase the value of n. This is achieved by designing carriers capable of being reversibly reduced/oxidised by multiple electrons per molecule. So called multi-electron charge carriers have a second advantage in that they can often be applied in symmetric systems, where the same charge carrier in different oxidation states is used in the positive and negative electrolyte. In the event of charge carrier transport through the membrane to the opposite half-cell (termed crossover), capacity fade is easily regenerated in symmetric systems as demonstrated in the all-vanadium RFB [12]. Conversely, asymmetric systems, which use a different charge carrier in the positive and negative electrolyte, often suffer from permanent capacity losses.

In the last two decades, research has shifted from metal-based charge carriers in aqueous solution to a new generation of charge carriers with tuneable physical and electrochemical properties to include inorganic, organic, and hybrid materials. In an effort to increase V_{cell}, there has been a growing interest in the development of non-aqueous electrolytes with wide windows of electrochemical stability [13]. Research has focused on designing charge carriers with high solubility in the chosen solvent, and rich electrochemistry at extreme potentials to maximise C, n and V_{cell} respectively. While each of these properties are paramount to enhance energy density, here, we focus on recent advances in the development of multi-electron charge carriers.

Figure 2 shows the components of a typical laboratory-scale RFB used for assessing the performance of charge carriers. Electrolyte is circulated via tubing between the reservoirs and electrochemical cell (typically powered by a peristaltic pump). Within the electrochemical cell, electrolyte is flowed through high surface area

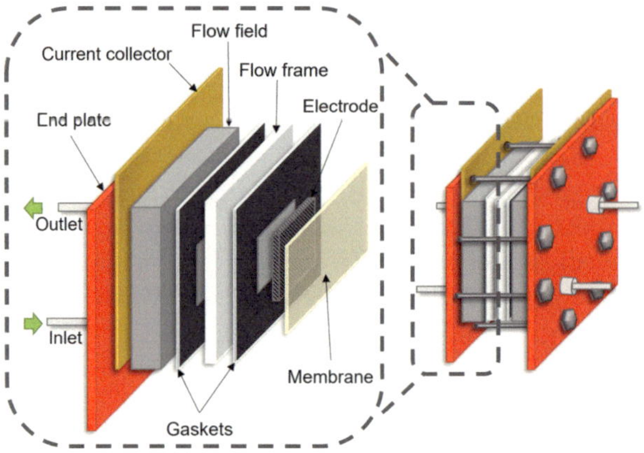

Figure 2.
Schematic of a typical laboratory-scale RFB used for assessing the performance of charge carriers. Battery components include the membrane/separator, electrodes, gaskets, flow frames, flow fields, current collectors, end plates, tubing, pump*, and electrolyte reservoirs*. * not pictured in schematic.*

electrodes where the redox reactions occur upon charge/discharge. The electrodes are typically composed of a carbon-based material such as graphite felt and are electrically connected to the current collectors and external circuit. The flow field (available in several configurations) ensures consistent flow of electrolyte to the porous electrode while minimising pressure drop across the cell.

The membrane/separtor divides the two half-cells and should be highly conductive, selective, and stable towards the electrolyte. High ionic conductivity is key to reduce ohmic resistance and thereby enable high power densities to be achieved. Membranes should allow transport of inert salts while preventing crossover of charge carriers, which can lead to capacity fade and reduced coulombic efficiency. Membranes/separators can be broadly classified as porous separators (separating based on size) and ion exchange membranes (separating based on charge). Identifying the most appropriate membrane/separator for novel RFB systems, where the chemistry is not fully understood, can be challenging. This is particularly true in the case of non-aqueous RFBs because very few commercially available membrane/separators have adequate performance in organic solvents [14]. Membranes present a barrier towards commercialisation for many next-generation RFBs since their inadequate performance reduces energy efficiency and they contribute up to 20% of the battery cost [15].

2. Organic charge carriers

There is a growing interest in the development of organic charge carriers as alternatives to metal-centred species, due to their tuneable properties and natural abundance of their elemental building blocks (C, H, N and O). Organic charge carriers investigated to date include nitroxide radicals such as 2,2,6,6-tetramethylpiperidinyloxyl (TEMPO), carbonyls such as fluorenone, benzophenone, phthalimides, quinones and anthraquinones, heterocyclic aromatics such as viologens, phenazines and phenothiazine, and cationic radicals such as dialkoxybenzenes and cyclopropenium, to name a few. **Figure 3** showcases a 'potential map' of organic charge carriers developed for next-generation RFBs in recent years. We direct the interested reader to consult review articles for further reading on organic charge carriers [13, 16–19].

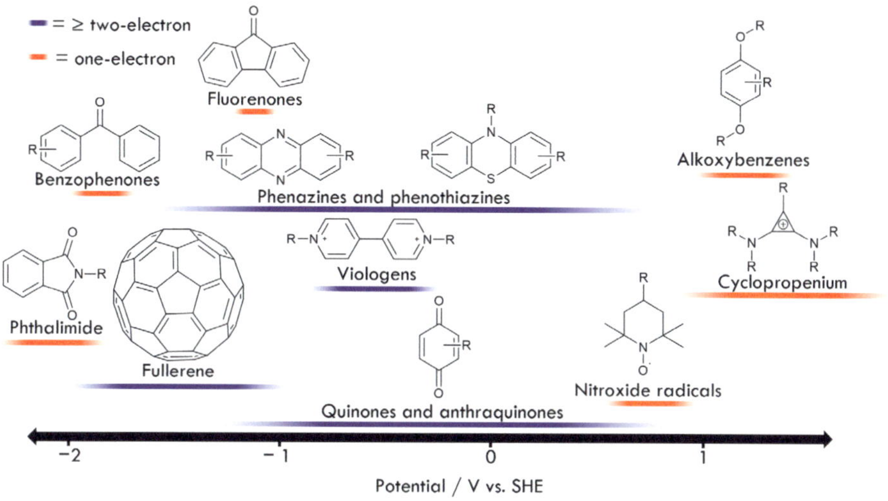

Figure 3.
Schematic showing the structure of organic charge carriers investigated for RFBs and the potential region in which they are redox-active. Orange bars indicate species that undergo one-electron redox reactions while those with blue bars are multi-electron charge carriers.

Several organic charge carriers such as viologens, quinones and phenazines, undergo two-electron redox reactions and have been investigated for RFBs. However, the π-conjugated cage structure of fullerene gives it uniquely rich electrochemical properties compared to other organic redox-active molecules. It can be reversibly reduced by up to six electrons at negative redox potentials, making it a desirable charge carrier for the negative electrolyte in RFBs. Without molecular modification, fullerene is nonpolar and unable to partake in hydrogen bonding, meaning its solubility in many solvents is poor [20]. Consequently, fullerene was first investigated as a charge carrier in RFBs as a bifunctional molecule where ferrocene groups (Fc) were covalently grafted to a C_{60} fullerene cage [21]. The functionalisation significantly enhanced the solubility of fullerene in ortho-dichlorobenzene (oDCB) from 0.037 M [22] to 0.12 M for the tetra-adduct of C_{60}Fc. The ferrocene groups served as the redox centre for the positive electrolyte, while the multi-electron redox processes of fullerene were accessed in the negative electrolyte. To balance the redox processes of fullerene, multiple ferrocene moieties were grafted to C_{60} (x = 1–4, where x indicates the number of ferrocene groups). The redox processes of the positive and negative electrolyte were separated by approximately 1.3 V and 1.8 V for the first and second reduction of C_{60} respectively. **Figure 4** shows the structure of the fullerene-ferrocene bifunctional charge carriers, termed C_{60}Fc, and the redox reactions occurring in the positive and negative electrolytes upon charge in the RFB.

The performance of C_{60}Fc charge carriers with x = 2–4 were investigated by galvanostatic cycling in coin cells. Charge carriers were assessed in symmetric and asymmetric configurations where indene-C_{60} bis-adduct was used as negative electrolyte. The coil cells were successfully cycled for 100 charge-discharge cycles but experienced considerable capacity fade which was attributed to three causes: (1) the low volume of electrolyte (~1 mL) in the coin cell assemblies meant that a significant proportion of capacity fade was attributed to electrolyte soaking into the absorbent glass fibre separator, (2) significant membrane crossover, which was alleviated by using the symmetric system rather than an analogous asymmetric system with indene-C_{60} bis-adduct negative electrolyte, (3) degradation of C_{60}Fc charge carriers,

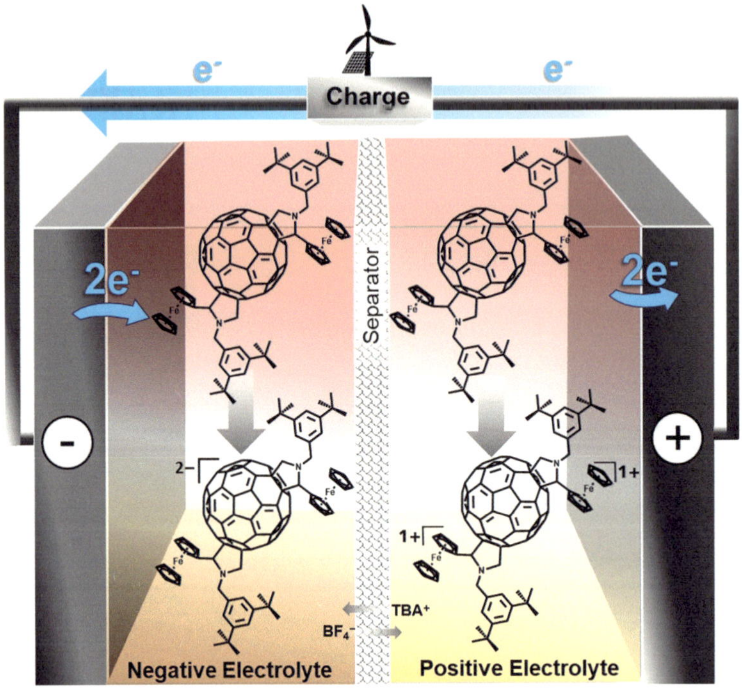

Figure 4.
Schematic showing the redox reactions of C_{60}Fc (bis-adduct) in a symmetric non-aqueous RFB. Upon charge, C_{60}Fc in the positive electrolyte undergoes a two-electron oxidation centred at the two ferrocene moieties. Simultaneously, C_{60}Fc in the negative electrolyte is reduced by two electrons at the fullerene core to a dianion. The reverse reactions occur upon discharge.

which was not explored in detail. Considering the rich electrochemistry and plentiful opportunities for functionalisation of fullerene, there is great scope for development of new fullerene-based charge carriers in the future.

The concept of combining two redox-active components into one bifunctional molecule was first demonstrated by Schubert and co-workers, who tethered TEMPO and phenazine moieties to a single molecule [23]. Upon charge, TEMPO was oxidised to an oxoammonium cation in the positive electrolyte while phenazine was simultaneously reduced to a dianion in the negative electrolyte. As shown in **Figure 5**, the bifunctional charge carrier contains two TEMPO groups (red) per phenazine (blue) in order to balance its two-electron redox chemistry. Both redox groups were covalently bonded via a water-soluble triethylene glycol linker (black), yielding a single charge carrier with V_{cell} of 1.2 V when applied in a symmetric RFB. The molecular engineering has a two-fold benefit in that it allows for its application in a symmetric system and enhances the saturation concentration in aqueous electrolyte beyond that of non-functionalised TEMPO and phenazine.

The advantage of symmetric systems in minimising capacity fade upon crossover can also be achieved by using a mixture of the positive and negative charge carriers in each half-cell. To justify the synthetic effort of bifunctional charge carriers, they must show advantageous properties compare to a mixture of the two individual redox-active molecules. In the two examples highlighted above, the bifunctional charge carriers have greater solubility than the individual redox-active groups and therefore a great theoretical energy density is achieved.

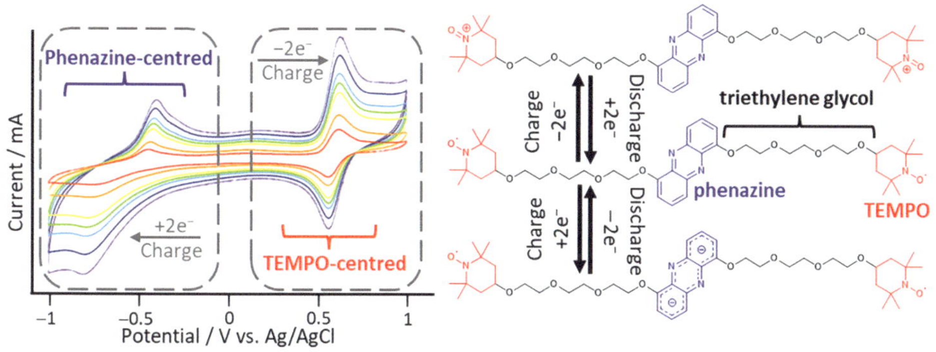

Figure 5.
Schematic showing the structure, redox reactions and cyclic voltammogram of a TEMPO-phenazine bifunctional charge carrier. Figure was adapted with permission form [21] copyright © 2016, American Chemical Society.

3. Coordination complex charge carriers

Transition metal coordination complexes are promising candidates for non-aqueous RFBs as they are often stable over multiple oxidation states and their properties are tuneable [24]. The first metal coordination complex to be investigated as charge carrier for non-aqueous RFBs was the ruthenium bipyridine (bpy) complex, $[Ru(bpy)_3]^{2+}$ [25]. Bpy ligands not only solubilise the metallic cation, they also provide additional redox activity to the molecule. In a symmetric system the Ru^{2+}/Ru^{3+} transition was targeted in the positive electrolyte while the bpy-centred reduction was targeted in the negative electrolyte. The redox processes were separated by 2.6 V, allowing for a high V_{cell}, but poor coulombic and voltage efficiencies were observed upon cycling. Within the last decade a range of metal-ligand combinations have been investigated to include nickel, cobalt, iron, vanadium, and chromium metal centres and acetylacetonate, terpyridine and dithiolene ligands to name a few [24]. Ligand design can have a remarkable impact of the charge carrier solubility, stability and redox properties and is therefore a key research focus for optimisation of metal coordination complexes. We direct the interested reader to two recent reviews for a more thorough examination of metal coordination complexes in RFBs [24, 26].

The chromium-centred bpy coordination complex, $[Cr(bpy)_3]^{3+}$ is of particular note due to its six one-electron reversible redox processes over a 2 V window (see **Figure 6**) [27]. The three most positive redox couples were attributed to the Cr^{3+}/Cr^{2+}, Cr^{2+}/Cr^{1+} and Cr^{1+}/Cr^{0} transitions, while the three most negative redox processes were ascribed to reduction of the three bpy ligands. The authors sought to enhance solubility in acetonitrile through ester-functionalisation of the bpy ligands. The complex functionalised with the most polar and flexible R group (2-(2-methoxyethoxy)ethyl) showed the most promising redox properties and solubility and was selected for battery testing. Galvanostatic cycling in a H-cell showed poor cycling stability when charged by three electrons but relatively stable performance when charged by two. The saturation concentration of both the neutral and 3+ complexes was tested and as anticipated the solubility was dramatically reduced in the neutral form. Despite the promising multi-electron redox properties of $[Cr(bpy)_3]^{3+}$, the poor solubility of the neutral complex (0.21 M in acetonitrile) and inadequate stability upon cycling, meant that the energy density of the system was limited to 10.2 Wh L^{-1}.

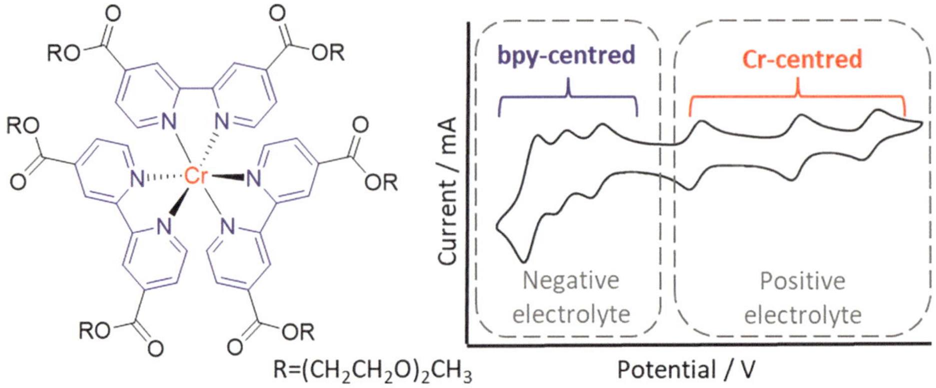

Figure 6.
Structure and cyclic voltammogram of functionalized $[Cr(bpy)_3]^{3+}$ coordination complex investigated for symmetric non-aqueous RFBs. Figure was adapted with permission form [25] copyright © 2015, American Chemical Society.

4. Polyoxometalate charge carriers

Polyoxometalates (POMs) are a class of discrete metal-oxide nano clusters composed of early transition metals (group 5 and 6) in their highest oxidation states. They can be represented by the general formula $[X_xM_mO_y]^{n-}$ where X is a hetero atom (usually P, Si, Ge, As) and M is a transition metal (typically V^{5+}, Mo^{6+} or W^{6+}). Their vast structural diversity, excellent stability and rich electrochemistry has seen their investigation for many energy conversion and storage technologies [28].

The earliest work on POM-based charge carriers for RFBs was conducted by Anderson and co-workers, who used the tri-vanadium substituted silicotungstate Keggin, $K_6H[SiV_3W_9O_{40}]$ (SiV_3W_9), in a symmetric aqueous system [29]. The POM undergoes a three-electron reduction centred at the vanadium metals and a further two, two-electron reduction processes centred at the tungsten metals. The vanadium-centred redox processes were separated from those of the tungsten by 0.8 V allowing for the application of SiV_3W_9 in a symmetric system. Prior to galvanostatic cycling, the charge carrier was reduced to $[SiV_3W_9O_{40}]^{13-}$ by bulk electrolysis to generate the fully charged negative electrolyte. Galvanostatic cycling in a 5 cm^2 RFB showed coulombic efficiency of >95% and modest capacity fade of <2% after 100 cycles (0.02% per cycle). Following 100 cycles, the electrolyte solution was recovered and used in a fresh cell with a new membrane. Full cell performance was restored suggesting that any capacity losses observed were not the result of POM degradation. Given the saturation concentration of SiV_3W_9 in water of 0.45 M, the three-electron redox reaction upon charge/discharge and V_{cell} of approximately 0.8 V, the theoretical volumetric energy density was calculated to be 14.5 Wh L^{-1}. While the energy density is half that of the all-vanadium RFB, this publication pioneered a new class of charge carrier with stable multi-electron redox processes.

Lu and Xiang extended the library of POM-based charge carriers to include the cobalt-centred tungstic acid, $H_6[CoW_{12}O_{40}]$ (CoW_{12}) [30]. Similarly to SiV_3W_9, the tungsten-centred redox processes of the POM were targeted in the negative electrolyte (two, two-electron reduction processes), while in this case, the one-electron oxidation of the central cobalt atom from 2+ to 3+ was targeted in the positive electrolyte. The separation of the cobalt- and tungsten-centred redox processes allowed for a larger V_{cell} of 1.25 V (0.8 V for SiV_3W_9). However, the positive electrolyte required four equivalents of CoW_{12} to balance the four-electron redox process occurring in the negative electrolyte. The laboratory-scale flow battery achieved coulombic efficiency >98% and good capacity retention over 30 cycles. Given the saturation concentration of CoW_{12} in water of 0.8 M, the theoretical volumetric energy density was calculated to be 16.8 Wh L^{-1} (accounting for necessary balance of POM equivalents in the positive and negative electrolyte).

Stimming and colleagues sought to maximise the rich electrochemistry of POMs by designing a asymmetric RFB with different POM-based charge carriers in the positive and negative electrolyte [31]. They investigated an asymmetric aqueous RFB using $[PV_{14}O_{42}]^{9-}$ (PV_{14}) and $[SiW_{12}O_{40}]^{4-}$ (SiW_{12}) as charge carriers for the positive and negative electrolyte respectively. PV_{14} is reversibly reduced by seven electrons in a single process at a relatively positive redox potential of 0.60 V vs. standard hydrogen electrode (SHE). The cyclic voltammogram of SiW_{12} has two reversible one-electron reduction processes with redox potentials of 0.01 V and − 0.21 V vs. SHE. SiW_{12} can be reduced by a further two-electrons but only the first two one-electron redox couples are accessible without significant hydrogen evolution. Consequently, a flow cell was assembled with two equivalents of SiW_{12} (n = 2) to balance the multi-electron redox process of PV_{14} (n = 4). Prior to galvanostatic cycling, PV_{14} was reduced by the addition of hydrazine to attain the discharged positive electrolyte. With an average V_{cell} of approximately 0.8 V and limited solubility of PV_{14} (demonstrated at 0.3 M), the theoretical energy density was calculated to be 13 Wh L^{-1}. Successful charge-discharge cycling was demonstrated in a 25 cm^2 flow cell and later upscaled to a 1400 cm^2 system, which remained stable of a 3 month period [32]. This provided a rare example of a next-generation RFB systems being tested at scale for extended periods of time. Capacity fade was attributed to the reoxidation of reduced POMs from trace oxygen which could be avoided with an airtight setup.

A step change in energy density came from the work of Cronin and co-workers where $Li_6[P_2W_{18}O_{40}]$ (P_2W_{18}) was reversibly reduced by 18 electrons in aqueous acidic conditions [33]. The authors found that the electrochemical properties of P_2W_{18} was highly dependent on pH and concentration and that reversible reduction by 18 electrons was only achievable at concentrations >100 mM under acidic conditions. Paired with HBr/Br_2 positive electrolyte, the asymmetric RFB demonstrated a V_{cell} of 1.25 V equating to a practical energy density of 225 Wh L^{-1}. Extrapolated to the saturation concentration of P_2W_{18} in water of 1.9 M, and assuming equivalent concentrations are achievable in the presence of supporting electrolyte, the authors calculate a theoretical energy density of >1000 Wh L^{-1}. It should be noted that while the saturation concentration of HBr/Br_2 is reported in the literature to be very high [34], the authors appear not to account for the volume of positive electrolyte in the energy density calculation.

RFBs based on aqueous electrolyte, such as those highlighted above, are limited to a maximum V_{cell} of *ca.* 1.5 V, beyond which, electrolysis of water occurs. Without targeted modification, the solubility of POMs in non-aqueous solvent is limited. In an effort to enhance the solubility of SiV_3W_9 in non-aqueous solvent, Anderson and co-workers conducted metathesis of the potassium counter cations to tetrabutylammonium (TBA) [29]. The TBA analogue of SiV_3W_9 was soluble in acetonitrile, propylene carbonate and methanol, but the redox processes became electrochemically irreversible in propylene carbonate, and was not explored further.

Barteau and colleagues were the first to report the application of POMs as charge carrier in non-aqueous RFBs [35]. They investigated the lithium salt of the Keggin phosphomolybdate, $Li_3[PMo_{12}O_{40}]$ (PMo_{12}), in acetonitrile with lithium trifluoromethanesulfonate (LiTf) supporting electrolyte. PMo_{12} undergoes two, one-electron quasi-reversible reductions centred at −0.21 V and − 0.57 V vs. Ag/Ag^+. For application in a symmetric system, the PMo_{12} electrolyte was first electrochemically reduced by one-electron to generate the discharged positive and negative electrolyte. Galvanostatic cycling was conducted with a one-electron redox reaction upon charge/discharge. As stated by the authors, this system does not exploit the full capabilities of POMs as multi-electron charge carriers nor the wide electrochemical stability window of non-aqueous solvents. Coulombic efficiency of 68% was achieved, which was substantially lower than that achieved for aqueous POM-based RFB. The low coulombic efficiency was attributed to crossover of the POM through the membrane, a common cause of inefficiency in non-aqueous RFBs. The saturation concentration of PMo_{12} in acetonitrile is relatively high at 0.8 M, but with only a one-electron redox reaction and V_{cell} of 0.35 V, the theoretical volumetric energy density was calculated to be 3.8 Wh L^{-1}.

While PMo_{12} can be reversibly reduced by two electrons in acetonitrile, the authors report the advantage of dimethylformamide (DMF) solvent in enhancing the electrochemical properties of the charge carrier [36]. In DMF, PMo_{12} can be reduced by an additional two electrons, enhancing the number of electrons transferred in the charge/discharge redox reaction to two and increasing V_{cell} to 0.45 V. The saturation concentration of PMo_{12} in DMF was reported to be 1.2 M, enhancing the theoretical energy density to 14.5 Wh L^{-1}. Barteau and co-workers later expanded the investigation to include an asymmetric RFB with PMo_{12} as charge carrier in the positive electrolyte and P_2W_{18} in the negative electrolyte [36]. The systems had a V_{cell} of 1.3 V and the number of electrons transferred upon charge/discharge increased to four, thereby further enhancing energy density.

The first example of significant molecular engineering of POMs to enhance solubility in non-aqueous solvent came from Matson and co-workers who investigated polyoxovanadate (POV) alkoxide clusters of the general formula $V_6O_7(OR)_{12}$ (where R = CH_3, C_2H_5) [37]. These materials display four one-electron redox couples over a potential range of 2 V, enabling their application in a symmetric system with two-electron transfer upon charge/discharge. Upon charge the POV undergoes a two-electron reduction at the negative electrode concurrently with a two-electron oxidation at the positive electrode. The research was extended to investigate alternative organic functionalisation of the POV surface. Introduction of a tridentate tris(hydroxymethyl)methane (TRIOL) ligand, increases solubility in acetonitrile to 0.6 M and retains the charge carriers cycling stability [38]. In a separate study, the solubility of POVs was increased by replacing several surface alkoxy groups with ethers, R = $C_2H_4OCH_3$, $C_2H_4OC_2H_5$ [39]. Clusters with a mixture of alkoxide and ether groups showed an impressive solubility of 1.2 M in 0.1 M [TBA][PF_6] in acetonitrile. While the increased solubility in organic solvent and multi-electron redox chemistry is promising for enhanced energy density, preliminary testing of the alkoxide-ether functionalised POVs in a laboratory-scale RFB showed steady capacity fade. Cyclic voltammetry of electrolytes following 30 cycles in a RFB indicated partial degradation of the POV clusters. For further reading on POV-based charge carriers for RFBs, we direct readers to a recent review article [40].

The concept of organofunctionalisation of POMs to enhance solubility in non-aqueous solvent was expanded in a recent publication, to include organic–inorganic hybrid POMs [41]. A phosphotungstate Keggin was hybridised with phenyl siloxane moieties to produce $TBA_3[PW_{11}O_{39}(SiC_6H_5)_2O)]$ ($PW_{11}SiPh$). Hybridisation enhanced solubility in acetonitrile by two orders of magnitude (0.6 M) compared to the parent POM (<1 mM). Similarly to POVs, $PW_{11}SiPh$ displays four one-electron redox couples over a potential range of 2 V. Prior to galvanostatic cycling the electrolyte was reduced by bulk electrolysis to attain the discharged positive and negative electrolyte for application in a symmetric system. Upon charge, $PW_{11}SiPh$ undergoes a two-electron reduction at the negative electrode concurrently with a two-electron oxidation at the positive electrode. The laboratory-scale RFB achieved high coulombic efficiency of >98% but capacity fade was observed. Similarly to Stimming and colleagues, the capacity fade was attributed to reoxidation of the reduced POM by trace oxygen in the electrolyte. Capacity fade was shown to be recoverable by bulk reduction of the electrolytes to the desired oxidation state. Organic-inorganic hybridisation is applicable to a broad range of POM geometries and elemental compositions, unlocking the possibility for the development of multi-electron charge carriers across a wide potential range (**Figure 7**).

Yan and colleagues explored the use of a sulphur-templated Wells-Dawson POM, $TBA_4[S_2W_{18}O_{62}]$ (S_2W_{18}), as charge carrier in both symmetric and asymmetric non-aqueous RFBs [42]. In the asymmetric system benzophenone was chosen as the negative electrolyte. Benzophenone undergoes a reversible one-electron reduction with redox potential of −1.75 V vs. $Ag^+|Ag$ (see **Figure 3**), while S_2W_{18} can be reversible reduced by one, one, then two electrons with redox potentials of 0.23, −0.15 and − 0.49 V vs. $Ag^+|Ag$. In the asymmetric RFB, four equivalents of benzophenone were used to balance the four-electron redox process of S_2W_{18}. Although not stated by the authors, reduction of S_2W_{18} to generate discharged positive electrolyte (or reduction of benzophenone to generate charged negative electrolyte) would have been necessary prior to galvanostatic cycling. The flow cell cycled successfully with V_{cell} of 1.54 V. Based on the saturation concentration of S_2W_{18} in acetonitrile of 0.11 M, the theoretical energy density was calculated to be 9.4 Wh L^{-1}.

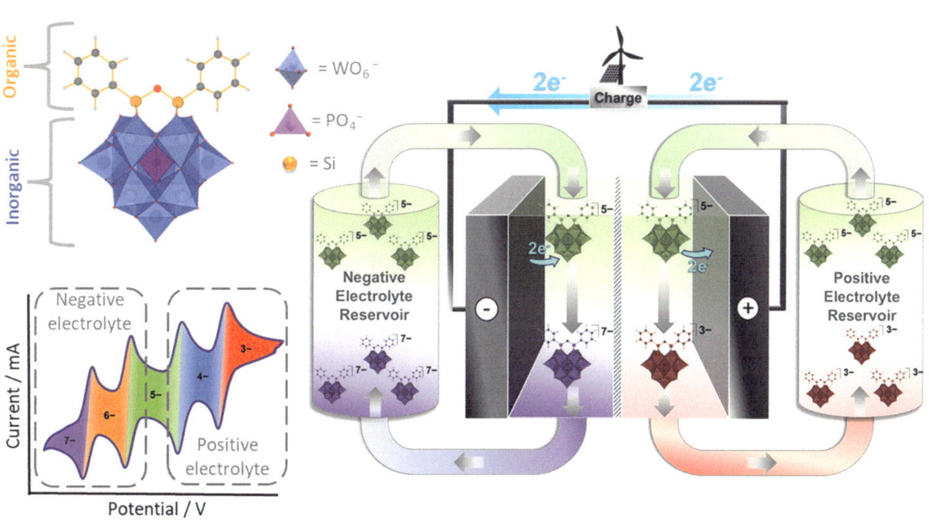

Figure 7.
Structure, cyclic voltammogram and RFB schematic of PW11SiPh, an organic-inorganic hybrid POM charge carrier. Figure was adapted with permission form [41] Copyright © 2021, American Chemical Society.

5. Conclusions

The use of multi-electron charge carriers is an effective approach to enhance the energy density of next-generation RFBs. Polyoxometalates stand out as a particularly promising class of materials due to their remarkably rich and reversible electrochemical properties. This is elegantly demonstrated by the 18-electron reversible reduction of P_2W_{18}, yielding an asymmetric RFB with practical energy density of 225 Wh L^{-1}. Organofunctionalisation of polyoxometalates is a valuable strategy to enhance solubility in non-aqueous solvent and for the tuning of redox properties and chemical stability. Other multi-electron charge carriers, such as metal-coordination complexes and bifunctional molecules, are realised through targeted molecular design and synthesis. Their rich electrochemical properties allow for their application in symmetric RFBs, thereby reducing the risk of capacity fade by membrane crossover. The bifunctional charge carriers also benefit from higher solubility than the isolated redox-active molecules.

Increasing the number of electrons transferred per molecule is a valuable strategy to enhance the energy density of RFBs. However, this parameter should not be targeted in isolation and should be considered alongside solubility, redox potential targeting, stability, cost and sustainability. In addition, the development of charge carriers for next-generation RFBs requires consideration of the flow cell assembly used for testing. Charge carrier performance depends on components such as membrane and tubing, and on the testing conditions such as flow rate, current density, and voltage thresholds. The lack of standardisation in testing conditions make it challenging to compare the performance of charge carriers.

The research reviewed here focuses on the development of novel charge carriers to enhance RFB performance. Most testing is conducted at low concentrations and in laboratory-scale RFBs. As the research matures, testing at scales more representative of the commercial product and detailed techno-economic analysis of the charge carrier-containing electrolyte will be required. Assessment of the costs of charge carriers, sustainability, safety, and practicality of synthesis will become increasingly important in the development of commercially viable next-generation RFBs.

Acknowledgements

The authors gratefully acknowledge the Engineering and Physical Sciences Research Council (EPSRC) for funding through the Centre for Doctoral Training in Sustainable Chemistry (EP/L015633/1). We also thank the University of Nottingham Propulsion Futures Beacon of Excellence for support.

Conflict of interest

There are no conflicts to declare.

Appendices and nomenclature

bpy	bipyridine
DMF	dimethylformamide
LIB	lithium-ion battery

POM	polyoxometalate
POV	polyoxovanadate
RFB	redox flow battery
TBA	tetrabutylammonium

Author details

Catherine L. Peake, Graham N. Newton* and Darren A. Walsh*
Nottingham Applied Materials and Interfaces (NAMI), The GSK Carbon Neutral Laboratories for Sustainable Chemistry, University of Nottingham, UK

*Address all correspondence to: graham.newton@nottingham.ac.uk and darren.walsh@nottingham.ac.uk

References

[1] Gür TM. Review of electrical energy storage technologies, materials and systems: Challenges and prospects for large-scale grid storage. Energy & Environmental Science. 2018;**11**(10):2696-2767

[2] Cameron JM, Holc C, Kibler AJ, Peake CL, Walsh DA, Newton GN, et al. Molecular redox species for next-generation batteries. Chemical Society Reviews. 2021;**50**:5863-5883

[3] Sánchez-Díez E, Ventosa E, Guarnieri M, Trovò A, Flox C, Marcilla R, et al. Redox flow batteries: Status and perspective towards sustainable stationary energy storage. Journal of Power Sources. 2021;**481**:228804

[4] Leung P, Li X, Ponce De León C, Berlouis L, Low CTJ, Walsh FC. Progress in redox flow batteries, remaining challenges and their applications in energy storage. RSC Advances. 2012;**2**(27):10125-10156

[5] Thaller L. Redox flow cell energy storage systems. In: Terestrial Energy Systems Conference Proceeding; Orlando, Florida, US. Cleveland, Ohio, USA: National Aeronautics and Space Administration; 1979

[6] Roe S, Menictas C, Skyllas-Kazacos M. A high energy density vanadium redox flow battery with 3 M vanadium electrolyte. Journal of the Electrochemical Society. 2016;**163**(1):A5023-A5028

[7] Zhao P, Zhang H, Zhou H, Chen J, Gao S, Yi B. Characteristics and performance of 10 kW class all-vanadium redox-flow battery stack. Journal of Power Sources. 2006;**162**(2):1416-1420

[8] Vanitec. Vanadium Redox Flow Battery Companies [Internet]. 2021. Available from: http://www.vanitec.org/vanadium-redox-flow-battery-vrfb-companies

[9] Dmello R, Milshtein JD, Brushett FR, Smith KC. Cost-driven materials selection criteria for redox flow battery electrolytes. Journal of Power Sources. 2016;**330**:261-272

[10] GRIDS: Grid-Scale Rampable Intermittent Dispatchable Storage. USA: US Department of Energy; 2010. Available from: www.osti.gov/biblio/1046668

[11] Trahey L, Brushett FR, Balsara NP, Ceder G, Cheng L, Chiang YM, et al. Energy storage emerging: A perspective from the Joint Center for Energy Storage Research. Proceedings of the National Academy of Sciences of the United States of America. 2020;**117**(23):12550-12557

[12] Lourenssen K, Williams J, Ahmadpour F, Clemmer R, Tasnim S. Vanadium redox flow batteries: A comprehensive review. J Energy Storage. 2019;**25**:100844

[13] Gong K, Fang Q, Gu S, Fong S, Li Y, Yan Y. Nonaqueous redox-flow batteries: Organic solvents, supporting electrolytes, and redox pairs. Energy Environmental Science. 2015;**8**(8):3515-3530

[14] Yuan J, Pan ZZ, Jin Y, Qiu Q, Zhang C, Zhao Y, et al. Membranes in non-aqueous redox flow battery: A review. Journal of Power Sources. 2021;**500**:229983

[15] Prifti H, Parasuraman A, Winardi S, Lim TM, Skyllas-Kazacos M. Membranes for redox flow battery applications. Membranes (Basel). 2012;**2**:275-306

[16] Wei X, Pan W, Duan W, Hollas A, Yang Z, Li B, et al. Materials and systems for organic redox flow batteries: Status and challenges. ACS Energy Letters. 2017;**2**:2187-2204

[17] Winsberg J, Hagemann T, Janoschka T, Hager MD, Schubert US. Redox-flow batteries: From metals to oRedox-active materials. Angewandte Chemie, International Edition. 2017;**56**(3):686-711

[18] Li M, Rhodes Z, Cabrera-Pardo JR, Minteer SD. Recent advancements in rational design of non-aqueous organic redox flow batteries. Sustainable Energy & Fuels. 2020;**4**(9):4370-4389

[19] Ding Y, Zhang C, Zhang L, Zhou Y, Yu G. Molecular engineering of organic electroactive materials for redox flow batteries. Chemical Society Reviews. 2018;**47**(1):69-103

[20] Sivaraman N, Dhamodaran R, Kaliappan I, Srinivasan TG, Rao PRV, Mathews CK. Solubility of C60 in organic solvents. The Journal of Organic Chemistry. 1992;**57**(22):6077-6079

[21] Friedl J, Lebedeva MA, Porfyrakis K, Stimming U, Chamberlain TW. All fullerene-based cells for non-aqueous redox flow batteries. Journal of the American Chemical Society. 2018;**140**:401-405

[22] Ruoff RS, Tse DS, Malhotra R, Lorents DC. Solubility of C60 in a variety of solvents. The Journal of Physical Chemistry. 1993;**97**(13):3379-3383

[23] Winsberg J, Stolze C, Muench S, Liedl F, Hager MD, Schubert US. TEMPO/phenazine combi-molecule: A redox-active material for symmetric aqueous redox-flow batteries. ACS Energy Letters. 2016;**1**(5):976-980

[24] Hogue RW, Toghill KE. Metal coordination complexes in nonaqueous redox flow batteries. Current Opinion in Electrochemistry. 2019;**18**:37-45

[25] Matsuda Y, Tanaka K, Okada M, Takasu Y, Morita M, Matsumura-Inoue T. A rechargeable redox battery utilizing ruthenium complexes with non-aqueous organic electrolyte. Journal of Applied Electrochemistry. 1988;**18**(6):909-914

[26] Palmer TC, Beamer A, Pitt A, Popov IA, Cammack CX, Pratt HD, et al. A comparative review of metal-based charge carriers in nonaqueous flow batteries. ChemSusChem. 2021;**14**(5):1214-1228

[27] Cabrera PJ, Yang X, Suttil JA, Hawthorne KL, Brooner REM, Sanford MS, et al. Complexes containing redox noninnocent ligands for symmetric, multielectron transfer nonaqueous redox flow batteries. Journal of Physical Chemistry C. 2015;**119**(28):15882-15889

[28] Li Q, Zhang L, Dai J, Tang H, Li Q, Xue H, et al. Polyoxometalate-based materials for advanced electrochemical energy conversion and storage. Chemical Engineering Journal. 2018;**351**:441-461

[29] Pratt HD, Hudak NS, Fang X, Anderson TM. A polyoxometalate flow battery. Journal of Power Sources. 2013;**236**:259-264

[30] Liu Y, Lu S, Wang H, Yang C, Su X, Xiang Y. An aqueous redox flow battery with a tungsten–cobalt heteropolyacid as the electrolyte for both the anode and cathode. Advanced Energy Materials. 2017;**7**(8):1601224

[31] Friedl J, Holland-Cunz MV, Cording F, Pfanschilling FL, Wills C, McFarlane W, et al. Asymmetric polyoxometalate electrolytes for advanced redox flow batteries. Energy & Environmental Science. 2018;**11**(10):3010-3018

[32] Friedl J, Pfanschilling FL, Holland-Cunz MV, Fleck R, Schricker B, Wolfschmidt H, et al. A polyoxometalate redox flow battery: Functionality and upscale. Clean Energy. 2019;**3**(4):278-287

[33] Chen JJ, Symes MD, Cronin L. Highly reduced and protonated aqueous solutions of [P2W18O62]6– for on-demand hydrogen generation and energy storage. Nature Chemistry. 2018;**10**(10):1042-1047

[34] Küttinger M, Wlodarczyk JK, Daubner D, Fischer P, Tübke J. High energy density electrolytes for H2/Br2 redox flow batteries, their polybromide composition and influence on battery cycling limits. RSC Advances. 2021;**11**(9):5218-5229

[35] Chen JJJ, Barteau MA. Molybdenum polyoxometalates as active species for energy storage in non-aqueous media. Journal of Energy Storage. 2017;**13**:255-261

[36] Cao Y, Chen JJJ, Barteau MA. Systematic approaches to improving the performance of polyoxometalates in non-aqueous redox flow batteries. Journal of Energy Chemistry. 2020;**50**:115-124

[37] Vangelder LE, Kosswatta arachchi AM, Forrestel PL, Cook TR, Matson EM. Polyoxovanadate-alkoxide clusters as multi-electron charge carriers for symmetric non-aqueous redox flow batteries. Chemical Science. 2018;**9**(6):1692-1699

[38] VanGelder LE, Petel BE, Nachtigall O, Martinez G, Brennessel WW, Matson EM. Organic functionalization of polyoxovanadate–alkoxide clusters: Improving the solubility of multimetallic charge carriers for nonaqueous redox flow batteries. ChemSusChem. 2018;**11**(23):4139-4149

[39] Vangelder LE, Pratt HD, Anderson TM, Matson EM. Surface functionalization of polyoxovanadium clusters: Generation of highly soluble charge carriers for nonaqueous energy storage. Chemical Communications. 2019;**55**(81):12247-12250

[40] VanGelder LE, Cook TR, Matson EM. Progress in the design of polyoxovanadate-alkoxides as charge carriers for nonaqueous redox flow batteries. Comments on Inorganic Chemistry. 2019;**39**(2):51-89

[41] Peake CL, Kibler AJ, Newton GN, Walsh DA. Organic–inorganic hybrid polyoxotungstates as configurable charge carriers for high energy redox flow batteries. ACS Applied Energy Mater. 2021;**4**(9):8765-8773

[42] Chen ZF, Yang YL, Zhang C, Liu SQ, Yan J. Manufacture of non-aqueous redox flow batteries using sulfate-templated Dawson-type polyoxometalate with improved performances. Journal of Energy Storage. 2021;**35**:102281